CONTINENTAL AND POSTMODERN PERSPECTIVES IN THE
PHILOSOPHY OF SCIENCE

This volume was born of a conference on postmodern philosophy of science offered at the InterUniversity Centre for PostGraduate Studies (IUC) in Dubrovnik in March, 1991 during what would be the final season of international conferences and courses.

A striking example of antemodern architectural harmony – its secular buildings grown out along sacred axes, its streets paved in the same fluid stone as the white walls framing the Adriatic sky – the city of Dubrovnik was shelled and the IUC villa destroyed by Serbian bombardment.

This collection is dedicated to the open spirit of multiculturalist collectivity and international scholarship of the IUC and to the people of Dubrovnik.

Continental and Postmodern Perspectives in the Philosophy of Science

Edited by
BABETTE E. BABICH
Department of Philosophy
Fordham University

DEBRA B. BERGOFFEN
Department of Philosophy and Religious Studies
George Mason University

SIMON V. GLYNN
Department of Philosophy
Florida Atlantic University

Avebury

Aldershot • Brookfield USA • Hong Kong • Singapore • Sydney

Published by
Avebury
Ashgate Publishing Limited
Gower House
Croft Road
Aldershot
Hants GU11 3HR
England

Ashgate Publishing Company
Old Post Road
Brookfield
Vermont 05036
USA

British Library Cataloguing in Publication Data

Continental and Postmodern Perspectives
in the Philosophy of Science
    I. Babich, Babette E.
    501

    ISBN 1 85972 192 3

Library of Congress Catalog Card Number: 95-78939

Printed and bound in Great Britain by
Ipswich Book Co. Ltd., Ipswich, Suffolk

Contents

III: ON APPLICATION: PRAXIS AND CRITIQUE

Contributors

Babette E. Babich, Department of Philosophy, Fordham University at Lincoln Center, 113 West 60th Street, New York, NY 10023, USA

Debra B. Bergoffen, Department of Philosophy and Religious Studies, George Mason University, Fairfax, Virginia, 22030, USA

Chip Colwell, Department of Philosophy, Villanova University, 800 Lancaster Avenue, Villanova, Pennsylvania, USA

Robert P. Crease, Department of Philosophy, State University of New York at Stony Brook, Stony Brook, New York, 11794, USA

Neil Gascoigne, Division of Philosophy, Anglia Polytechnica University, East Road, Cambridge, CB1 1 PT, England

Simon V. Glynn, Department of Philosophy, Florida Atlantic University, PO Box 3091, Boca Raton, Florida, 33431, USA

Charles Harvey, Department of Philosophy, University of Central Arkansas, Conway, Arkansas, 72035, USA

Patrick A. Heelan, William Gaston Professor of Philosophy, Department of Philosophy, Georgetown University, Washington, DC, 20007, USA

Ladelle McWhorter, Department of Philosophy, University of Richmond, Richmond, Virginia, 23173, USA

Felix O'Murchadha, Philosophische Fakultät, Universität Wuppertal, 42329 Wuppertal, Germany

Alphonso Lingis, Department of Philosophy, Pennsylvania State University, University Park, Pennsylvania, 16802, USA

Brian Pronger, School of Physical and Health Education, University of Toronto, 320 Huron Street, Toronto, Ontario, M5S 1 A1, Canada

Daniel Rothbart, Department of Philosophy and Religious Studies, George Mason University, Fairfax, Virginia, 22030, USA

Raphael Sassower, Department of Philosophy, University of Colorado, Colorado Springs, CO, 80933, USA

Acknowledgements

Thanks are due to Patrick A. Heelan, Executive Vice-President at Georgetown University, and formerly Dean at the State University of New York at Stony Brook, for his support of the conference, *Postmodern Prospects for a Transformed Philosophy of Science*, Dubrovnik, 18-22 March 1991. Thanks are also due to Georgetown University for financial assistance toward the publication of this volume.

Original co-directors of the Dubrovnik conference were Babette E. Babich, (Organization), Nicholas Davey, Patrick A. Heelan, Holger Schmid, and Thomas Seebohm.

On the Idea of Continental and Postmodern Perspectives in the Philosophy of Science

Babette E. Babich, Debra B. Bergoffen, and Simon V. Glynn

Introduction

Hermeneutic, phenomenological, genealogical and postmodern critiques of science may be conceived as a radicalization of those contemporary analyses of science which take their point of departure from the fundamental principle of complementarity and recognize that science can never be a mirror of nature; that there are no neutral observers; that all experiments are theory-laden; that there are no simple facts. These perspectives sensitize us to the historical, political, social, and cultural dimensions of science. They force us to revisit the epistemological claims of science and insist that we ask whether and to what extent the idea of scientific privilege can be sustained.

As post-metaphysical, the hermeneutic core of postmodernism sets itself the task of interpreting discourses and narratives. When the discourse is modern science, postmodern and continental style philosophy poses such questions as: what is the source of the power of this discourse? what is the meaning of the world provided by this discourse? what are the moral and political implications of this discourse? Given its focus on interpretation and rhetoric and its rejection of the modern distinction between the rational and the irrational, postmodernism treats the sciences as embedded in, related to, and – running up against the modern ideal of clarity and distinctness – as ineluctably *contaminated* by other cultural languages and practices. While some might say that this perspective negates the possibility of science, others insist that such a postmodern view allows us to understand crucial discourse and practice relationships, that a specifically continental and postmodern perspective gives us a better understanding of the hows, whats, and whys of science.

With this understanding, the modern idea of truth as reflective of nature gives way to postmodern (Nietzschean) questions of interpretation, valuation, and perspectivalism. The modern idea that the conflict of interpretations can be mediated or resolved in such a way as to provide a single coherent theory which corresponds to the way things are, gives way to the thought of an infinitely interpretable reality where diverse, divergent, complementary, contradictory, and incommensurable interpretations contest each other without, however, canceling each other out. That the traditional idea of science cannot hold in these circumstances is clear. What we explore here is the extent to which these circumstances preclude the idea of science *per se*. Toward this end, the following essays review the relationship between postmodernism and traditional and continental philosophies of science by examining scientific methods and disciplines, the histories of the sciences, the place of science within the modern world, the value accorded to science and the epistemology of the scientific project.

Risks: Postmodern, continental approaches to the philosophy of science

To juxtapose postmodern and continental philosophical thought with the routinely analytic (and roundly modern) discipline of the philosophy of science is a chancy thing. Continental style philosophy is far from recognized as a viable approach to the philosophy of science[1] and the flip contentiousness seemingly constitutive of postmodern thought is, if anything, even less appropriate for the philosophy of science.[2]

Yet it is not true that there has never been any invocation of postmodernism and its categories within traditional (read: analytic) approaches to the philosophy of science. Stephen Toulmin, one of the foremost "forecasters" of the philosophy of science, was one of the first to write on "postmodern" science and philosophy (Toulmin, 1985) and his recent *Cosmopolis* (Toulmin, 1990) offers a gentle version of postmodern critique as it poses (and proposes an answer to) the question of the "Hidden Agenda of Modernity."[3]

Toulmin's perspective is sagely optimistic where he offers an assessment of the current (so-called postmodern) circumstance of modernity. This optimism is for the most part mirrored in the present collection. For the present authors, the postmodern condition represents not so much an established paradigmatic reversal of the modern, its difference from the plainly modern signified for example by appropriate double-coding, playfulness, pluralism, etc., as a condition of modernity as it is still in need of clarification and above all as a condition calling for recognition. Thus the postmodern condition is understood as a project proposed for reflection (or "thought") concerning just where it is that

2

contemporary thinkers find themselves, to use Toulmin's words, with respect to "practical philosophy, multidisciplinary sciences, and transnational or subnational institutions." As such a reflective orientation, the postmodern prospect is inherently, perhaps necessarily ambivalent. For Toulmin, this ambivalence reflects two contrasting dispositions proposed as alternate responses to the contemporary condition: imagination and nostalgia. Charged by imagination, we may welcome the postmodern prospect as one "that offers new possibilities, but demands novel ideas and more adaptive institutions; and we may see this transition as a reason for hope." Or else, as Toulmin's alternative would have it, we are remanded to the fearful nostalgia of passivity and impotence, turning "our backs on the promises of the new period, in trepidation, hoping that the modes of life and thought typical of the age of stability and nationhood may survive at least for our own lifetimes." (Toulmin, 1990, 203)

Rather than representing a premodern or romantic reactionary spirit, as Jürgen Habermas and other critics of postmodern notions argue, the essays to follow do look hopefully forward. Yet it must be with *both* hope and trepidation (a quintessentially postmodern combination) that what follows is an ironicised critique of the prototypically modern project of the philosophical understanding as well as of the professional practice of science. Such critiques, such skepticism and irony, are inevitable. For while a latent nostalgia is not the watchword of this collection, as Freud, Nietzsche, and recent world events remind us, hope is not without anxiety (where whatever first calls for hope is sparked and defined by the threat of dis- and misappointments). Even where hope is justified, transitions never go smoothly. In this dissonance, a continuing shock to the seamlessly modern progress-ideal, the postmodern condition ultimately calls for nothing less elusive than Nietzschean "light feet."[4]

The essays: structure and overview

This collection is formally postmodern in three respects. There is no one definition of either science or the postmodern; no single answer to the question of the relationship between science, the philosophy of science, and postmodernism. The modern rationalistic axiom that postmodernism, science, and philosophy of science are fundamentally incompatible is set aside. The issue of the association of postmodern critique and the philosophy of science is framed as a possibility, not eliminated in advance.

The constellation and composition of the present collection raises the question of boundaries – once again in object fashion. In asking about the prospect of a postmodern philosophy of science, these essays seek to explore the extent to which a critical (philosophical) perspective not originating in the sciences but rather in the cultural spheres of art and the humanities can be meaningfully applied to the theoretical and practical sciences. Further, consistent with the spirit of postmodernism, the configuration of this collection evokes and underlines suspicions concerning its own project. If it is the case that the boundaries between the humanities and the natural and social sciences are quasi arbitrary marks of power, might it also be the case that the move to elude and collapse these boundaries marks another power play? Is it a power play of philosophy which without directly empowering philosophy as a bastion of truth moves to regain philosophy's erstwhile position as queen of the sciences by dismantling those domains of knowledge and power which have succeeded in overshadowing it? If this last question is not the immediate subject of this volume it remains at the margin – important for postmodern thinking where the margin counts as much as anything at the center.

Whether or not we agree to call it postmodern, we can agree that we are living in a multi-national, capitalist, nuclear world society conditioned throughout by science and technology. This imperative condition of late, post, or third-stage modernity requires our attention, however we define ourselves theoretically. For modernists cannot ignore the changed and changing circumstances of enlightenment rationality or the scientific project. And postmodernists cannot ignore the question of science and the ways it is being (and might be) practiced in a postmodern world.

The first section, *Postmodern Continental: Propædeutic and Parody*, explores the relationship between postmodernism and the philosophy of science together with a provocative or polemical critique of analytic styles in philosophy to outline some of the disputed issues between postmodern, continental and modern, analytic philosophies of science.

A defender of the postmodernist position, *Raphael Sassower* argues that philosophers of science and postmodernists are often unaware of one another. Arguing that this ignorance should be remedied, Sassower offers to introduce them to each other and suggests that their awareness of each other would produce a more radical critique of science than that offered by such philosophers of science as Popper, Feyerabend, Kuhn, or Polanyi, as well as a more relevant critique of the contemporary situation than that of the postmodernists grounded in literature, architecture, or aesthetics. Distinguishing the postmodern from the pseudo-liberal critique of science, Sassower sees the work of Donna Haraway as an important link between the postmodern and feminist critiques of science and as countering the charge that postmodernism is relativist and irresponsible.

Babette E. Babich takes polemical issue with the traditional definition of the philosophy of science as such, suggesting that a genuine philosophy of science should critique rather than precommit itself to accepting and adopting science's epistemic assumptions and methods, as is the received practice and ideal of

4

analytic style philosophy of science. Using direct (argument) as well as indirect (parodic) means, Babich challenges the unilateral conviction and coherence of the analytic style in the philosophy of science. Only by drawing from a broad range of alternative perspectives, especially those deriving from the continental tradition and, indeed, ranging beyond the assumptions of the postmodern perspective, is a critique and understanding of many of the otherwise taken for granted claims, methods, and thetic presuppositions of science possible. Only such an approach can be accounted an authentic philosophy of science.

The next section, *Theory of Science: Quantum Deconstruction*, specifically addresses intersections between hermeneutics and phenomenology and the philosophy of the natural and human sciences. These essays articulate the hermeneutic, historical, and social dimensions of the natural sciences, analyze the role of metaphor and analogy in the history and practice of science, and suggest that a postmodern perspective would resolve many of the so called paradoxes of contemporary science, to bridge the gap between traditionally styled or analytic essays in the philosophy of science and the broadly continental project of the present collection.

Patrick A. Heelan's challenging conception of an "anti-epistemology" argues that the quantum theory needs to be given what he names an ontological rather than an epistemological interpretation. An epistemological interpretation is one related to cognitive content while an ontological interpretation is relative to the activity of representing the cognitive content. These theories describe phenomena as revealed through socio-historical processes of empirical inquiry by local communities of expert witnesses rather than as objective realities. Such theories imply a role in the scientific account for two non-classical freedoms, i.e., for social factors and for history. The ontological viewpoint here proposed is inspired by traditions as old as Aristotle and as new as Heidegger and has the further postmodern virtue of using quantum theory to elaborate an interpretive account of objectivity applicable to the social as well as the physical sciences.

Robert P. Crease explores the analogy between the conceptual prestructuring or interpretation of experience, and the scripting or scoring of a theatrical or musical performance, to further suggest the affinity of nature and culture. Like the symbols on a musical score, which have a relationship both to the notes that are played, and to the other symbols (notes) scored, scientific theories relate both to the world and to each other. The former relationship, Crease suggests, is the focus of the experimentalist, describing the "performances" of the facts, while the latter is the focus of the theoretician, who is concerned with consistency between theories. Moreover, just as we cannot simultaneously observe every aspect or element of an historically and socio-culturally located performance, so scientific performances or experiments are similarly located and presented perspectivally.

5

Simon V. Glynn traces the route from the phenomenological reduction, via Heidegger's ontological hermeneutics to the deconstruction of dualistic epistemologies and the concomitant demise of correspondence theories of truth and veridicality. He finds a parallel route, from the reduction of supposedly experience-independent, intrinsic identities of objects, to empirical properties which vary with context, and which may therefore be deconstructed into systems of extrinsic or structural relations. This, Glynn points out, amounts to the demise of absolute, non-relational, identity. Showing how the application of such a postmodern epistemology to Einstein's Relativity Theory, Heisenberg's Uncertainty Principle, Quantum Field Theory, and Niels Bohr's theory of Complementarity, dissolves many of the paradoxes associated therewith, Glynn suggests its further application to more recent paradoxes in physics.

Expanding on the conceptual underpinnings of observation, *Daniel Rothbart* examines how analogical models are pivotal to the observation of phenomena. As is clear from the practice of conceiving electricity as analogous to a fluid, or referring to the information "processing" capacities of the mind, or to light waves, analogies enable science to render comprehensible the ostensibly incoherent patterns of nature. So when an experimenter "reads" tracks in a cloud chamber as the passage of small particles, such a reading depends upon the analogical projection of patterns from familiar symmetries. Rothbart concludes that the human intervention that characterizes every level of scientific access to nature's secrets includes the creative discovery of powerful analogies in nature.

Charles Harvey argues that the reductionistic meta-narrative ideals of totalization inveterate to both the natural and the human sciences are conceptually feasible and performatively demonstrative. Nevertheless, due to the sense-parameters distinguishing "the natural" from "the human," the two types of sciences can never be semantically conjoined. Yet, Harvey also argues, in a postmodern world, we are best off letting the sciences methodologically totalize, while teaching ourselves a phenomenological calm about life-wordly sense-gambits, a learning project which might be balanced with an existential agility about what we count as real, and when we do so.

The ultimate section, *On Application: Praxis and Critique*, presses this last postmodern challenge to the modern demarcations used to distinguish the natural and social sciences from each other as well as from other interpretive strategies. The authors question the role of metaphor in science and ask about the relationship between science and its environs. Specifically attending to controversial issues, such as sexism, racism, AIDS, and, most radically, the connection between science and *Eros/Thanatos*, these essays shift the focus from the natural or physical sciences to the social and human sciences.

Debra B. Bergoffen draws our attention to the ambiguity of the body of knowledge metaphor. Though usually understood with reference to the object

6

of scientific discourse (the mysterious feminine body that must be tortured to reveal her secrets) it can also refer to the knowledge produced by science. Influenced by Nietzsche and Lacan, Bergoffen asks: How shall we understand science's promise to provide us with a reliable body of knowledge? Pursuing this question allows her to decipher our understandings of the object and project of science and to challenge the demand for a unified body of knowledge. Attending to what Lacan has taught us about the powers of the imaginary, Bergoffen alerts us to the Nietzschean possibilities of a "Gay Science," – a science that recognizes the needs for coherence, consistency, and unity as it celebrates the heterogeneity of the given and pursues the fluidity of the lived body.

Taking up the question of the relation between scientific knowledge and power, *Chip Colwell* focuses on Foucault's analysis of the relation between the medical narrative and economic, social, and political institutions. Public support for the sick lead to their removal from home into hospitals or clinics where the patient could more readily be objectified as the site of symptoms and where previously localized medical discourses became the grand narrative of medicine. The clinician's control over the body of the diseased meant that Death, traditionally the point at which the secrets of the disease were irrevocably lost, became, via autopsy, the point of revelation. Like other forms of knowledge, medical knowledge is mediated by economic, political, and social institutions, and thus by the relations of power, which it therefore reflects, and to which in turn it contributes. Colwell suggests that the postmodern doctor would recognize the grand medical narrative in all its importance and in its objectifying tendencies as only one among a number of possible discourses on disease.

Ladelle McWhorter explicates Foucault's notion of power as a non-reified process or event, emanating from many points, and at least as capable of generating institutions and the relations between them – and thus institutionally constituted notions of truth and knowledge for instance – as of being a reflection of them. Attempting to demonstrate that Foucault's analysis may be extended beyond the social sciences to the natural sciences, she turns to biology as a case in point. Arguing that the account given by biologists of the relations between species and races provided justification for slavery, taboos against interracial marriages, etc., McWhorter claims that such classifications consolidate, extend, and reflect the interests of those at the center of power.

Following a Heideggerian reading, *Felix O'Murchadha* points out that our knowledge of the world implies a view of and therefore a relation to the world and that we are constituted by those very perspectives and acts of interpretation by which we come to what we name the truth. The cultural and the natural world are mediated by the same conceptual or symbolic systems and hermeneutic interpretations that constitute the human as such. In consequence, the concerns

7

of epistemology (theoretical knowledge) and practical existence (ethics) are pragmatically united; our knowledge of the world is inherently ethical.

Radicalizing this last point of inquiry, *Neil Gascoigne* asks whether we can make sense of the self-conscious, ethically responsible, subject if the subject/object dichotomy has been deconstructed. When postmodernists, as well as some modernists, reject the noumenal or transcendental self that Kant identifies as the free center of a universal ethics, they leave us with the empirical self, as a reified ego unable to escape causal determinism, and thereby incapable of assuming ethical responsibility for its actions. Arguing that we can distinguish what is represented from how it is represented, and that such a distinction may leave room for an ethically responsible subject in spite of its misrepresentation as a reified ego, Gascoigne notes that such a solution is not without difficulties. It is for instance, sometimes claimed (Freud, Marx) that the outside observer is in a better position than the actors themselves to comprehend the true meaning and significance of their actions, a claim which in effect reinstates the absolutist perspective of a transcendental signifier or noumenal I.

Alphonso Lingis notes that Martin Heidegger set out to bring to light the history of the specific form of the technological imperative at work in our theoretical and practical reason. But Lingis argues that Heidegger's account does not sufficiently distinguish what is specific to the diverse ordinances that command our perception, our technology, and our social fields as the lived world and body. The representations science constructs of the perceptual field, the technological field, and the social field are not continuous with one another.

As an illustration of Lingis's emphasis on the ordinance of the lived body and its environmental referentiality, the dialectic tension between disembodied and embodied reification is offered in a precisely physicalistic context by *Brian Pronger*. In the final essay of this volume. Pronger examines the tendency to reify reflective or analytic distinctions and finds this correlative with the tendency to objectify or hypostatize the self. Pronger identifies this tendency with *Thanatos* which he contrasts with *Eros*: the erotic urge to the synthetic unity of being in the process of lived becoming. While modern science is, from Pronger's point of view, clearly in the service of *Thanatos*, a postmodern science would acknowledge, along with *Eros*, the unity and thus the essential relatedness of all elements of existence.

References

Babich, B.E. (1994), "Philosophy of Science and the Politics of Style: Beyond Making Sense," *New Political Science: A Journal of Politics and Culture*, Summer/Fall 1994: 30/31, pp. 99-114.

Babich, B. E. (1993), "Continental Philosophy of Science: Mach, Duhem, and Bachelard," in Kearney, R. (ed.), *Routledge History of Philosophy: Volume VIII*, Routledge, London, pp. 175-221.

Baudrillard, J. (1992), "The Ecstasy of Communication," in Jencks, C. (ed.), *The Postmodern Reader*, Academy Editions, London, pp. 151-157.

Bergoffen, D. (1990), "Nietzsche's Madman: Perspectivism without Nihilism," in Koelb, C. (ed.), *Nietzsche as Postmodernist. Essays Pro and Contra*, State University of New York Press, Albany, pp. 57-71.

Bohm, D. (1992), "Postmodern Science and a Postmodern World," in Jencks, C. (ed.), *The Postmodern Reader*, Academy Editions, London, pp. 383-391. Also in Griffin, D.R. (ed.), *The Reenchantment of Science: Postmodern Proposals*, State University of New York Press, pp. 57-68.

Bohm, D. (1981), *Wholeness and the Implicate Order*, Routledge, London.

Borgmann, A. (1992), *Crossing the Postmodern Divide*, University of Chicago Press, Chicago.

Eco, U. (1984), *Postscript to The Name of the Rose*, Weaver, W. (trans.), Harcourt, Brace Jovanovich, New York.

Glynn, S.V. (ed.), (1986), *European Philosophy and the Human Social Sciences*, Gower, Aldershot.

Griffin, D.R. (1988), "The Reenchantment of Science," in Griffin, (ed.), *The Reenchantment of Science*, State University of New York Press, New York, p. 1-56. Abridged in Jencks, *The Postmodern Reader*, pp. 354-72.

Lyotard, J.-F. (1979), *La Condition Postmoderne: Rapport sur le savoir*, Les Editions de Minuit, Paris. (1984) Bennington, G. and Massumi, B. (trans.), University of Minnesota Press, Minneapolis.

Toulmin, S. (1990), *Cosmopolis: The Hidden Agenda of Modernity*, University of Chicago Press, Chicago.

Toulmin, S. (1985), "Pluralism and Responsibility in Post-Modern Science," *Science, Technology & Human Values*, 10:28-37.

Rosenau, P. M. (1992), *Postmodernism and the Social Sciences*, Princeton University Press, Princeton.

Simpson, L. C. (1995), *Technology, Time, and the Conversations of Modernity*, Routledge, New York.

Notes

1. This is not because there are no proponents of such approaches to the philosophy of science but because these approaches go unrecognized: presumably unread and quite crucially uncited. See however the first named editor's essays, Babich (1994), (1993), and below.

2.   Postmodernism hardly enters mainline or traditional analytic style philosophy although it figures in the socio-political axis in positive reference made by such continental authors as Jean Baudrillard, Umberto Eco, and Jean-François Lyotard. Popular science authors like the more religiously minded physicists such as, for example, the late David Bohm, tend to use the term *postmodern* fairly freely. For further discussion of the alliance between religion and science, especially ecology, see Griffin. In philosophy of technology there is some evidence of a serious reception of the concept of the postmodern condition, particularly with regard to its multiculturalist and feminist dimensionality (see, among others, Borgmann and Simpson). In the social sciences invocations of the postmodern are common enough to be featured in titles (e.g., Rosenau: *Postmodernism and the Social Sciences*). It is of course telling, as a failure to which the present volume is addressed, that discussions of the postmodern, be it condition or quandary, are not featured in contemporary philosophy of science with its dominant focus on natural science.

3.   Toulmin himself, although a master of the analytic art of non-citation, names Frederick Ferré the "pioneer" of postmodernity in the natural sciences but adds "see also the final essays in Stephen Toulmin, *The Return to Cosmology*." Toulmin, 1990, p. 213.

4.   See on this, the first two named editors' contributions to Clayton Koelb's *Nietzsche as Postmodernist*, particularly Debra Bergoffen's "Nietzsche's Madman: Perspectivism Without Nihilism."

I
POSTMODERN CONTINENTAL: PROPÆDEUTIC AND PARODY

1 Prolegomena to Postmodern Philosophy of Science

Raphael Sassower

Prologue

The journey to the land of postmodernity, a space often misunderstood and ridiculed, a terrain too fuzzy to be of use for topographers, is undertaken partly with Karl Popper, joining Paul Feyerabend for some length, and then seeking the company of feminists. All of these companions contribute to a sensitivity – in the sense of concretizing epistemological debates and suggesting the need for a psychological setting different from that of the positivists and for a politics different from that of the elitists – of postmodern philosophy as applied to science, so as to become a postmodern philosophy of science, even when the specific concerns of the leading postmodern writers are different from those of philosophers of science and vice versa. The two parties ignore each other. I wish to engage them both – hopefully to acquaint them with each other, well in accord with their respective theories though regrettably not yet exemplified in their practice.

For example, Popper is the great developer of the last major attempt to demarcate science from all other forms of discourse and Feyerabend criticizes this attempt. Now the "problem of demarcation" (no matter how it is narrated) is in fact a political problem, one that bespeaks a form of privileging – a "metanarrative" (Lyotard, 1984, xxiii-xxv) – that the Lyotardian "postmodern condition" refuses to accept in principle. Popper's demarcation forms a certain boundary between "fiction" and "science" and that boundary translates into a fence of an exclusive club, into a status symbol of positions of power, and as such it is an obstacle on the road of critical analyses of the scientific culture in which we live (Ormiston & Sassower, 1989, Ch. 2) which Popper himself advocates. This situation overlooks some of the very issues raised by Popper himself.

For example, Popper claims that metaphysics does matter in science, that Truth must be relegated to the position of an honorary concept in the scientific community but one without any political clout, much like the Queen of England, and that the unresolved problem of philosophy of science remains psychology: how do scientists come to their new ideas? What disposition must one have to be able to challenge established models and propose alternative ones? But still, how does psychology fit into the context of scientific discourse?

The answer in short is offered with the aid of tools from the discourses of postmodernists. In some respects I find solace in the psychological possibilities that open up within the critical discourses of postmodernity, in the sense of empowering all narratives and reorganizing institutional power relations. (Sassower) Of course one could say that Michael Polanyi (1958) and Thomas Kuhn (1970) have given us as much as we need to know about the workings of what they refer to as the "scientific community," analyzing issues of leadership and responsibility, raising questions about the ways in which the methodology of science can and should be pushed further along the paths of a research program, with or without the tampering that can be had with its metaphysical "hard core." (Lakatos)

For my taste that is not enough. I recommend radicalizing the critiques of science in the wake of postmodern literature which however, regrettably for its part, overlooks too many aspects of the scientific enterprise. The radicalization of the critiques of science I recommend has to do with the political character that has been detected in the domain classically renowned for its alleged value-neutrality. This means, e.g., that when Karl Marx attacks classical political economists he simultaneously criticizes the theoretical content of their models as well as their political conclusions, not to mention their methodological foundation. As such, he invokes the connection between "theory" and "practice." Contemporary feminists critique the sciences in a similar fashion, locating scientific research within its socio-economic and political context. I end my exploration with Donna Haraway, whose social-feminist views on science are essential for the appreciation of what should count as postmodern philosophy of science with its variations and differences as contrasted, for example, with some pseudo-liberal ideals expressed by Richard Rorty.

Background

"Postmodernity" has suffered suspicion and ridicule since its most recent incarnation in literary and philosophical circles. Perhaps most noticeable has

14

been the refusal to grant postmodernism any status whatsoever, that is, the refusal to acknowledge its very existence as a mode of thinking and doing if not an intellectual movement. For example, C. Barry Chabot claims:

> (1) that no satisfactory and widely accepted account of postmodernism now exists; (2) that much of what is called postmodern in fact derives directly from modernism; and (3) that most arguments for its existence achieve their initial plausibility largely through improvised characterizations of modernism, especially characterizations that neglect its nature as a second-order concept. (Hoesterey, 37)

Others, like Martin Jay, have warned against "a leap into the postmodernist dark," (Hoesterey 108), as if there is something dangerous about the psychological settings and possibilities offered by postmodernism or as if postmodernism would return us to the "dark ages."

One way of understanding the psychological settings of postmodernism is by emphasizing that broadening the horizon of possibilities – of narrative production and personal exchange – does not entail forming hierarchies of choices. (Sassower & Ogaz) As Matei Calinescu says: "What aesthetic dualism or pluralism shows is simply that a choice does not necessarily imply a summary dismissal or ignorance of other available alternatives." (Hoesterey, 167) There is a difference between the openness of postmodernism and the openness of "objectivity" and rationality as the guiding principles of "modernity." The openness of postmodernity attempts to blur disciplinary boundaries and encourages the violation of traditional rules or the use of experimental techniques that juxtapose theories and practices usually kept at arm's length from each other (because they are either incommensurable or historically differentiated).

Postmodernism illustrates that the articulation of different forms of discursive exchange by itself does not necessarily enhance the possibility of greater engagement across intellectual barriers, an engagement whose political potential has been sought all along by postmodernism. At the same time, "Science" – that category of modernity presumed to be fighting the prejudices of church authorities and the superstitions of the uneducated – has lost its leverage as "peace maker;" it is perceived to be a category of power and (white male, hegemonic, etc.) domination/oppression that eschews debate when these debates are deemed too critical (and therefore threatening to its authority) or not critical enough (and therefore not worthy of attention at all). In attempting to engage postmodernism with science I try to bring together the potential each offers separately in order to have a greater impact on critical discourse. I will return to the question of critical and radical engagement of postmodernism and science

15

after examining briefly what may be considered the tenets of postmodern philosophy of science, clarifying along the way why I find it important to focus on science and its attendant critiques.

Should there be postmodern philosophy of science?

The question, Should there be postmodern philosophy of science? addresses two related contentions: (1) postmodernism does not differ from modernism (as quoted above), and (2) if there is a difference, there has always been a postmodern philosophy of science. Indeed, if one were to agree that what characterizes postmodernism is its psychological settings and possibilities, then it is possible to illustrate that postmodernist attitudes, such as a recognition of the interpretive activity required in "explaining" or writing about "nature," have been present in/with many bona fide scientists; for example, Max Planck in the twentieth century, as Valerie Greenberg (1990) shows in relation to problems raised by Franz Kafka. Going back to the seventeenth century, Francis Bacon's sensitivity to the interpretive mode of scientific research is easily detectable in his *New Organon* (Ormiston & Sassower, 27-29) and as such can be viewed as an "adherent" of postmodernism. Richard Rorty, for instance, would have to agree with this contention since he too sees Bacon closely aligned with Feyerabend as a "prophet of self-assertion," as opposed to aligning him with the prophets of "self-grounding," Descartes and Kant. (Hoesterey, 92)

Methodologically, Rorty's comment on Bacon can be understood as advocating a view of philosophy of science that focuses primarily on questions of orientation towards the material under study. According to this view, epistemological questions are more directly connected to psychological dispositions and the fashions one adopts both historically and socially in terms of the environment in which the research is undertaken. So, whether one agrees that a postmodernist flavor has accompanied even the most rigid modernist versions of scientific inquiry or that it is a relatively new/different kind or classification of/in philosophy of science, it seems that the examination of postmodern philosophy of science has to do with a constructivist view of the scientific enterprise. The construction of science, what Stephen Toulmin describes as the shift from "scientist-as-spectator" to "participant," (Toulmin, 29) becomes at once politicized undertaking, one that can no longer shield itself from public engagement because of its presumed value-neutrality.

Despite the cavalier use of the term "postmodern philosophy of science," one would be hard-pressed to find a clearly defined area of research that fits this term or a consensus among its advocates. Currently there are several areas of study that come under the label of philosophy of science: history of science, methodology of science, sociology of science, even science, technology, and society. But the category of postmodern philosophy of science – combining two disputed discursive spaces, postmodernism and philosophy of science – is at once intriguing and suspect. What does this discursive labyrinth in fact stand for? Is it a chronological or historical designation that demarcates between modern and *post*modern philosophy of science? Does it herald a "new" way of thinking about and doing science? Besides, is the focus science or the philosophy of science?

Having stated that a postmodern orientation can be traced to a number of historical texts, it is nevertheless possible to detect a difference between so-called modernist and postmodernist versions of philosophy of science. At the same time, the novelty of the approach is not in terms of "originality," but rather in terms of difference. The reflexive conception of postmodernists that their efforts are different is a direct critique of previous or other attempts, such as the Enlightenment modernists who claim superiority (of approach, methodology, and success) over previous practices (relegated at that point in time to nothing more than superstition and wishful, unscientific thinking). The jargon of difference – whether in Derrida's (1988) or Lyotard's sense (1988) – tries to ameliorate previous contentions of superiority by opening fields of research, offering additional options that in their introduction do not claim either superiority or the necessity to do away with all previously known forms of discourse. This is what is at stake when speaking of displacement as opposed to replacement (e.g., Ormiston & Sassower). Moreover, the jargon of difference ensures, as Toulmin concedes, that it is no longer necessary to attempt a "positivist" reduction of scientific research to a "single set of methods." (Toulmin, 29)

But so far I have said little about what postmodern philosophy of science *is*. Perhaps it is wise to take Susan Rubin Suleiman's advice and problematize this very question: "If postmodernist practice in the arts has provoked controversy and debate, it is because of what it 'does' (or does not do), not because of what it 'is'." (Hoesterey, 113) Having changed the question from what it is to what it does, Suleiman continues to explain that postmodernist "doing" is invoked "in a particular place, for a particular public . . . at a particular time." (Hoesterey, 119) The contextualization of postmodern practices, as Charles Jencks explains, differentiates them from others: "Modernists and late-modernists tend to emphasize technical and economic solutions to problems, whereas postmodernists tend to emphasize contextual and cultural additions to their inventions." (Hoesterey, 9)

Coming up with a more precise definition is quite difficult, because definitions tend to be too broad or too narrow, and then they need to be qualified forever, just as Kuhn still has to do with his notion of "paradigm." Moreover,

17

providing a definition goes against the postmodern grain since it pretends to capture a moment that it too fleeting to catch. Instead of definition, then, let me follow Suleiman's lead and add that postmodernist practices – those of philosophy of science included – belie and illustrate an orientation, an approach, an attitude. Whose attitude?

Being playful, Umberto Eco claims in an interview with Stefano Ross that one defines the postmodern nowadays as "everything the speaker approves of;" but he also acknowledges that postmodernism is "a spiritual category . . . a world view," one that is characterized by "irony" and a "metalinguistic play." (Hoesterey, 242-243) Closer to the terminology I have used so far, Eco suggests that he "would consider postmodern the orientation of anyone who has learned the lesson of Foucault, i.e., that power is not something unitary that exists outside of us." (Hoesterey, 244) Perhaps it is important to note in this context that Eco wishes, along postmodernist lines of orientation, to insist that "our behavior in the world ought to be not *rational* but *reasonable* . . ." (Hoesterey, 244) – transcribing the epistemological discourse of science and technology into that of psychology and sociology.

Back to the specific orientation I wish to ascribe to or express through postmodern philosophy of science. What I have in mind here is an orientation that lies somewhere between Paul Feyerabend's anarchism of "anything goes" and Karl Popper's rationalism of falsification. Incidentally, my appeal to these thinkers differs from the appeal that Gerard Radnitzky, for instance, has to Popper and Wittgenstein as the two polar opposites that "carve" the landscape of twentieth century philosophy of science.(Radnitzky) For my purposes, the company of these two thinkers betrays my own genealogy and does not set "limits" to what can and should be considered under the ambiguous title of postmodern philosophy of science or postmodern science.

In its provocation to proliferate a multiplicity of scientific models and methods of inquiry, postmodernism adheres to a kind of pluralism envisioned by Feyerabend. In both cases, there is a flexible attitude towards the latest theory and scientific gimmick: they are welcome newcomers as long as the "old guard" is not dismissed or demolished. This is the critical engagement I have in mind, perceived as a process of displacement – where the "new" and the "old" continue to confront each other – as opposed to a process of replacement – where the newcomers do away with any traces of the old guard. There is even some affinity on the question of relativism in both Feyerabend and postmodern conceptions: the criteria of evaluation of models depend on their theoretical framings which also need to be contextualized culturally. Feyerabend pushes methodological debates to their logical extremes and cracks them open by inviting any method of inquiry to participate and be accorded similar (scientific) legitimacy as that of already "established" ones. Acknowledging Feyerabend's

18

logical indebtedness to Popper's views and with a tint of irony I ask: what can Popper add to this openness?

Popper adds restraint to Feyerabend's presumed recklessness. The Popperian restraint does not limit potential "newcomers," for he solicits any and all hypotheses to be considered scientific as long as they refrain from shielding themselves from critical empirical testing. That is, no matter how wild a speculation ("conjecture"), it should be permitted to confront the scientific "court of appeals" and have its day in court. If it is falsifiable in principle, then it retains its scientific status until it is in fact falsified and a revision is proposed (see Popper, 1959 and also 1963). The Popperian restraint, one that can also be found in Feyerabend regardless of some protestations to the contrary, maintains that critical examinations be based on both rationality and empirical data.

Is such a restraint too harsh? Will it rob postmodern philosophy of science of its French mystique and potential? The answers to these questions depend on the method of application of the restraint. Enlightenment adherents of the scientific method went to an extreme to ensure a strict application of reason and rationality and codified their applications in strict logical terms. By the same token (and with a great deal of overgeneralization), empiricists appealed to the empirical data in a rigid and faithful manner that overlooked some subtleties, such as the appreciation of theory-laden observations in Hanson's sense. (Hanson, 1958) According to Toulmin, what "went wrong" in the modernist project as it progressed into the present is not realizing that "the planetary system is a quite exceptional system. No other system of entities in the natural world lends itself to prediction in the same way." (Toulmin, 30) There is a way to be rational without being narrowly committed to a specific set of methods or logical principles, just as there is a way of being empirical without becoming an empiricist. A balancing act of this sort may have informed Eco in setting the preference for being "reasonable" rather than "rational."

The appeal of postmodern philosophy of science lies in its elasticity, its flexible adoption of the ideas proposed by Popper and Feyerabend together as well as those proposed from literary criticism's quarters, architecture, and art criticism. I will continue to focus at present on Popper and Feyerabend, for the juxtaposition of their ideas may show a discursive intersection attributable to postmodern philosophy of science, one that is echoed in feminist critiques of science (more below). Feyerabend's principle is considered irrational, for he provides no strict criteria of demarcation between science and pseudo-science, thereby legitimating ipso facto any theory and model (like Chinese medicine as opposed to Western medicine) that works under certain conditions as scientific. (Feyerabend, Ch. 18) By contrast, Popper's principles are considered rational and his criteria of demarcation are considered strong contenders to the logical positivist standards of this century. So, how does postmodern philosophy of science wedge itself between rational and irrational principles? Can it disregard this binary, and if yes, at what price? That is, is the postmodernist elasticity not a vice rather than a virtue, allowing contradictions to coexist?

It seems that the postmodern elasticity works simultaneously on several levels of investigation and practice. First, since postmodernism refuses to accept any principles whatsoever as setting a permanent foundation for anything, the adaptation of principles deemed both rational and irrational is unproblematic. That is, without a firm foundation, the foundation one uses is context-bound or "situational" and therefore quite "shifty," it changes every time one uses it. As such it becomes temporal and bound by specific "pragmatic" goals in the light of which it is applied. (Ormiston & Sassower, 17, 19) On this issue there is a growing feminist literature that shifts the focus of American pragmatism to a different discourse.

Second, even the anarchistic principle associated with Feyerabend is not presented in an irrational form; on the contrary, there is a long and arduous rational discourse that argues for the acceptance of this viewpoint and appeals to notions of consistency and non-contradiction. In this sense, then, one can appreciate a minimalist use of logical principles, no longer as strict rules of practice but rather as guidelines that enable the discourse to "get off the ground." Once again, the attitude here is not one that disallows certain vocabularies or forms of discourse nor one that requires à la Habermas a "consensus;" instead it is a liberating attitude, one that attempts to explain away traditional binaries by setting a very low "threshold" over which anyone can pass. (This attitude can be found, for example, in the works of Agassi and Jarvie.)

Third, once critical evaluation is viewed as a tool for the improvement of general discourse, be it scientific or not, the quest for scientific legitimacy on grounds other than critical evaluation becomes superfluous if not mere appeal to authority. Critical evaluation, as the lowest possible threshold of communication, turns out to be the reference point according to which the rules of a critique are formed. Lowering standards is objectionable; as Clement Greenberg says: "Modernism, insofar as it consists in the upholding of the highest standards, survives – survives in the face of this new rationalization for the lowering of standards." (Hoesterey, 49)

The elasticity of postmodern philosophy of science continues to occasion numerous criticisms and objections, some of which were already mentioned. Reiterating several concerns at this juncture would be helpful in order to appreciate some of the issues raised and mentioned below by feminist critiques. One objection to the elasticity portrayed by anything associated with the term postmodernism denounces its relativism. Relativism is anathema to scientific inquiry, for it would presumably allow two competing and incommensurable theories to coexist without the ability to differentiate their scientific validity and credibility (e.g., Laudan, 1990). If science has made any contributions, these contributions have been commonly associated with the ability to distinguish between belief and truth, between trust in someone's statements and evidence that these statements ought to be trusted, between the imagined and the real. For example, a standard comparison is between magic and science, two areas of research and practice that are supposedly clearly distinguishable using the standard modernist criteria of demarcation between science and pseudo- or non-science. Once relativism is adopted, the criteria for such a distinction are lost, thereby suspending the prestige attributed to science and technology. As Charles Newman claims: "Post-Modernism is an ahistorical rebellion without heroes against a blindly innovative information society." (Newman, 10)

Another and related objection has to do with the vacuous nature of postmodernism when one tries to apply it to political practice or technological experimentation. Having robbed science and technology of their esteemed status and materiality – data are not "real" but constructed – and having opened their quarters to anyone and anything that wishes to be called science and technology would result in total chaos and the loss of possible appeals to stability, security, and even common sense. Science has been traditionally the refuge of brave intellects from religious and superstitious speculation and persecution (claiming its victims along the way, like Giordano Bruno in 1600). Postmodernist views of science demolish this last refuge, leaving nothing, not even common sense, protected from power plays and abuse of authority (e.g., Foucault). One expression of this problem is well summarized by Felix Guattari (1986) who voices his concern with the politically and materially vacuous structuralist analysis perpetuated by postmodernists/poststructuralists, an analysis that yields no platform or agenda for change. As such, this analysis is considered irresponsible.

From science to politics through critique

Postmodern philosophy of science is a political mode of discourse when it tries to apply its epistemological concerns to specific situations in which the scientific community is involved. A postmodern orientation to the study of science may have little to do with challenging, for example, current views on quantum mechanics. Yet, in its insistence on continued forms of critiques, it can encourage continued criticism of the construction and acceptance of the latest "received view." The criticism could come from the meekest voice and from the most remote of locations; yet this would possibly be criticism that displaces one theoretical framework with another, not merely supplements it. It may seem by now, then, that all that is needed is some level of tolerance, and that postmodernism contributes to make this possibility come true. However, as will become clear below, the appeal to (liberal) tolerance loses its force as one evaluates the impact of different critiques and the competition for pronouncing the "final word" or "verdict" on the future of philosophy of science, philosophical inquiries in general, or the enterprise called science. In this respect, then, tolerance may be a necessary but definitely not a sufficient condition for the development of critiques of science.

Late twentieth-century critiques of science diverge in their methodological focus and political goals. Among them one can count sociological critiques (e.g., Latour & Woolgar), those of Marxists (Aronowitz) and feminists (e.g., Harding), and even those of the philosophical establishment, from Popper to Kuhn. As we approach the next century, different intellectual scenarios can be envisioned in light of these critiques. First, each mode of critique will continue its efforts to undermine traditional presuppositions concerning science without regards to other critiques. As such, each effort will ignore deliberately those of others. Second, noticing the efforts of others, each mode of critique will attempt to ascend to a position of hegemony from which all others are perceived as either contributive or incidental. Eventually, one mode will dominate and overshadow all others, in the sense of excluding all others from making separate contributions. Third, the rivalry among different modes of critique will cause a disarray of efforts so that their effects will be marginal or overlooked systematically.

The three scenarios may occur simultaneously, so that no clear picture of the intellectual scene will be possible to ascertain. Yet, under all three scenarios, one possible outcome looms heavily: the ineffectiveness of any and all modes of critique of science to change anything so that the status quo is maintained. A plea for some engagement among the different modes of critique is a political plea (in the academic and intellectual contexts) to preserve the possibility of changing not only specific practices of scientists but also the entire political structure of the capitalist environment in which they operate. One possible site for such an engagement may be postmodern philosophy of science. With this in mind, I would like to see more clearly how alliances are implemented between Marxists, feminists, postmodernists, Popperians, and any other group of critics that is committed to radicalize the practices – linguistic and institutional – of scientists.

If the "landscape" portrayed above makes sense, that is, if the three scenarios could possibly take place, and if the plea for critical engagement on strategic grounds is deemed worthy, then it is useful to pay special attention to the form of feminist critique enunciated with the work of Donna Haraway. Her work illustrates how one can at once employ both "external" and "internal" critiques, the former associated primarily with philosophers, historians, and sociologists of science and the latter associated with scientists themselves. The dialectical twist undertaken by feminists (along Marxist lines) conflates these two classifications of the notion of critique (internal and external) and claims that it is essential for a critique to be at once immanent to the practices and theories of science as well as be "outside" in order to have a perspective that may be blurred from the "inside," one that may transcend traditional, oppositional binaries.

A dialectical critique attempts to understand the internal structure of a theory or model or mode of behavior. This understanding – a deconstruction in contemporary parlance – sets the stage for a reconstruction or reconfiguration that may be pushed simultaneously in two directions. On the one hand it allows an opportunity to examine some implicit or tacit presuppositions in an explicit fashion, as for example, the feminist focus on gender-based forms of discourse and practice; on the other hand it suggests ways in which the theory or model or mode of behavior can be pushed to their "logical extremes" in order to evaluate their potential collapse (as in the case of the capitalist mode of production in Marxist terms).

Once undergone the critical process of evaluation, there is also the possibility of setting-up alternatives to the theory or model or mode of behavior. Historical examples abound. One example encompasses the debates between the inductivists among the Vienna Circle members and Karl Popper. The inductivist/verificationist agenda was criticized for its inappropriate demarcation of science from non-science. Falsification was supposed to be the guiding principle for testing empirically the scientific "content" of conjectures and hypotheses. Did the Popperian principle of falsification (with its attendant notion of verisimilitude) *replace* the inductivist principle of empirical confirmation? Does his critique infuse a discussion of the metaphysical commitments of scientists – e.g., a belief that the Truth can be revealed once and for all? Or, as some critics continue to claim, has the Popperian critique been usurped by the probabilistic inductivist framework and is no longer an alternative?

Another example is Kuhn's view of paradigm shifts and the relation between normal and revolutionary science: does it *replace* traditional views of the methodology of science (theories devoted to a cumulative and continuous growth of scientific knowledge)? Kuhn's ideas are supported by his interpretation and reconstruction of the historical record, as well as by his focus on certain sociological aspects that organize and influence the growth of scientific knowledge within the scientific community. Kuhn's critique, then, attempts to recast science in a different light, one presumably deemed more accurate or appropriate. But Kuhn's critical success has become, almost immediately with its introduction, a target of relentless critiques: some charging his views with the irresponsibility that accompanies irrationalism, some claiming that his view is so vague and ambiguous that it is no coherent critique at all.

Marxist critiques of science, from Karl Marx to Stanley Aronowitz, share the common thrust that in their articulation they expose science for what it really is: a capitalist and bourgeois activity whose goal is the exploitation of the working (and probably by now also the middle) class. This critique of science tries to denounce the mystification surrounding scientific jargon and texts in order to reveal the ideals of truth to which Marxists adhere. Contemporary Marxists critiques are at times lumped together with feminists critiques (from Keller to Harding and Haraway), though each uses different techniques of critique and has different objectives in mind. Overlooking for the moment the great divergence of techniques and ideals, it is clear from these critiques that they wish to replace scientific practices prevalent today with some other, better forms of discourse and practice.

Are postmodernist critiques different? With a bit of irony, the answer is at once yes and no. Just as Marxist critiques all hold on to banners, such as "exploitation" and "alienation" as a useful terms with which to anchor their discourses, so do postmodern critiques wave among many other banners that of "displacement." This term is supposed to replace (with a twist of irony) the term "replacement," so that the postmodern discourse should be accorded residency alongside or next-door, without thereby pretending to surpass or overcome previous discourses in a Hegelian sense of *Aufhebung*.

As someone who is not "of" the feminist discourse, but nevertheless regards himself a decent listener, I find interesting parallels and points of intersection between feminist and postmodernist critiques of science. But that is not enough. I remarked in my Prologue above that feminist critiques are "essential" in order to make sense of postmodern critiques of science, yet one may recall Suleiman's warning about the apparent "opportunism" that may undermine an effort such as mine in this context. That is, one may dismiss my attempt to "listen" to feminist critiques as a fetishized attempt to hook unto the latest intellectual fad or fashion. Yet, as Suleiman remains us, "There is, I believe, an element of mutual opportunism in the alliance of feminists and postmodernists, but it is not necessarily a bad thing." (Hoesterey, 116) If I understand her correctly, Suleiman acknowledges the possibility of forming strategic alliances that would have a stronger political force to change the "established" or canonical view of science and technology.

It is not only "interesting" for postmodern philosophers of science to consult the work of feminists, such as Haraway, but necessary. For it seems that the canonical spokespeople for postmodernism (e.g., Derrida or Deleuze and even Lyotard) shy away from considering science and technology in the manner that Haraway does. Or when they in fact mention science and technology, their critiques fall short of providing the sort of analysis that they provide regarding architecture or literature (e.g., Lyotard maintains a certain level of deference to scientific "descriptions," see Sassower & Ogaz). Feminist critiques, therefore, lead the way for an improved version of a postmodern critique of science. Without them postmodern philosophers of science will resort to their old examples, just as I have done above, concerning the Vienna Circle and Popper or Kuhn and Feyerabend.

Feminist and postmodern engagement

Haraway confronts discursive practices in an explicit political fashion: "The evidence is building of a need for a theory of 'difference' whose geometries, paradigms, and logics break out of binaries, dialectics, and nature/culture models of any kind. Otherwise, threes will always reduce to twos, which quickly become lonely ones in the vanguard. And no one learns to count to four. These things matter politically." (Haraway, 1991, 129) For Haraway, the allure of postmodern "pluralism" is not reduced to an engagement in the creation of different vocabularies for the sake of it. For example, when she says: "Splitting, not being, is the privileged image for feminist epistemologies of scientific knowledge. 'Splitting' in this context should be about heterogeneous multiplicities that are simultaneously necessary and incapable of being squashed into isomorphic slots or cumulative lists," (Haraway, 1991, 193) she is concerned with a political situation. In fact, she translates the postmodern concern to pluralize epistemology into "an argument for *pleasure* in the confusion of boundaries and for *responsibility* in their construction." (Haraway, 1991, 150)

Whereas postmodernists have been accused of being "irrational" or irresponsible, advocating what seems to be intellectual "chaos," feminists, such as Haraway, insist on the responsible construction of blurred boundaries in the aftermath of deconstruction. In Haraway's construction, then, "Some differences are playful; some are poles of world historical systems of domination. 'Epistemology' is about knowing the difference." (Haraway, 1991, 161) Epistemology, according to her, takes on a different form of orientation, one that no longer merely "plays" with the latest data or reconstructs a new paradigm that would revolutionize scientific thought, but instead contextualizes critical judgments and affirms particular political commitments concerning any contestant for the status of paradigm.

Offering a different vocabulary may be deemed problematic if one understands that there is a difference between replacement and displacement of one vocabulary with another. Haraway does not wait for another to recognize this problem, but problematizes it herself: "The feminist dream of a common language, like all dreams for a perfectly true language, of perfectly faithful naming of experience, is a totalizing and imperialist one." (Haraway, 1991, 173) In order to avoid becoming one more "stage" in some universal Hegelian system, she recommends perceiving the feminist critique of science as engaged in "cyborg politics," one that is about "noise." (Haraway, 1991, 177) Her insistence on noise as opposed to a unified, clear, and totalizing language would ensure that a feminist critique is not about anti-science that rebuilds dualisms. (Haraway, 1991, 181) She reminds her readers that "Science has been utopian and visionary from the start; that is one reason why 'we' need it." (Haraway, 1991, 192)

When Haraway explains her epistemological - read political - viewpoint, she enters the debate about "Situated Knowledge," one that promises to supersede binaries. (Haraway, 1991, 187) According to her, "Situated knowledges require that the object of knowledge be pictured as an actor and agent, not a screen or a ground or a resource, never finally as slave to the master that closes off the dialectic in his unique agency and authorship of 'objective' knowledge." (Haraway, 1991, 198) What differentiates her posture on this question from Baconian and Kantian attempts to appreciate objective knowledge in intersubjective terms (as was the accepted norm of the Royal Society of London) or from the attempts of pragmatists to contextualize knowledge in practical, goal-oriented frameworks is her sense of responsibility. She says:

> The moral is simple: only partial perspective promises objective vision. This is an objective vision that initiates, rather than closes off, the problem of responsibility for the generativity of all visual practices . . . Feminist objectivity is about limited location and situated knowledge, not about transcendence and splitting of subject and object. In this way we might become answerable for what we learn to see. (Haraway, 1991, 190)

The answerability of feminist critiques rests on a redefined notion of rationality: "The science question in feminism is about objectivity as positioned rationality." (Haraway, 1991, 196) "Positioned rationality" could be equated with the critical and relativized rationality about which Agassi and Jarvie speak or with the one associated with postmodern forms of contextualization (Ormiston & Sassower). But, as was already mentioned in the case of postmodernism in general, can Haraway's view escape the standard criticisms that it falls into a form of irrationalism or relativism? Her response is: "The alternative to relativism is partial, locatable, critical knowledges sustaining the possibility of webs of connections called solidarity in politics and shared conversations in epistemology." (Haraway, 1991, 191)

Liberal intolerance and radical engagement

"Webs of connections" has been understood also as "labyrinths," wherein it is difficult to distinguish one narrative from another, since each is linked in more than one way to every other. (Ormiston & Sassower)   Systematically deconstructing any claim for singularity and uniqueness may pose a problem, as Haraway recognizes.   So, it seems that she prefers to use the notions of "solidarity" and "conversations" to explain her general orientation.

The "S" word (solidarity) has been used extensively by Richard Rorty.   I mentioned Rorty above as a fellow-interpreter of Bacon's work in a postmodernist mold.   But Rorty the postmodernist turns out to be a liberal of a particular kind, perhaps bourgeois, perhaps privileged.   For example, he speaks of "human solidarity" (Rorty, 1989, xv) – understood in terms of a "liberal utopia" (Rorty, 1989, 190) – in his interpretation of the postmodern orientation that includes "contingency" and "irony."   But Rorty's solidarity is not Haraway's, and this difference is important to note if we are to understand the critical engagement between postmodernism and science that is no recurrence of liberal modernist engagements of science.

Though Rorty is quick to invoke the notion of utopia in his work, it is clear it is not the utopian vision of Marx or that of Haraway.   For him the utopian dream of the "liberal West" is of "Tolerance rather than that of Emancipation." (Rorty, 1991, 213)   Rorty's disdain for postmodern politics can be surmised from the following statements.   Revolutionary politics, for Rorty, is "no more than intellectual exhibitionism", an activity that is "not interested enough in building causeways" to connect intellectual "islets" with the "mainland." (Rorty, 1991, 221)

Rorty positions himself *vis-à-vis* French philosophers and does not directly address in this context feminist critiques of science.   But if the alliance I try to establish between postmodern and feminist critiques of science can come to fruition even temporarily, then Rorty's posture would undermine it at once, antagonizing feminists along the way.   That is, Rorty's views are relevant in this context for he can easily divert any effort to radicalize the discourses of science and bring them to a liberal fold that fails to see the stakes feminists, for instance, have in critically engaging science and its practitioners.   Most problematic in Rorty's assessment of the status of postmodern and feminist critiques is his cavalier use of the term "mainland."   The solidarity he seeks is with the mainland, that is, the liberal intellectual establishment, one that is comfortable, as he says, with "splitting the difference between Habermas and Lyotard, of having things both ways." (Hoesterey, 94)   But some differences are incommensurable, and sometimes differences cannot be simply "split": the truth may not lie somewhere in the "middle."

Haraway's notion of "conversation," (see also Haraway, 1991, 201) parallels my sense of a critical or radical engagement, where disagreement is not glossed over nor necessarily mitigated. Postmodern philosophy of science can be the forum for such an engagement, where critical evaluations and radical utopias would be solicited continuously. My view differs from Rorty's since it no longer expects that conflicts need to have "causeways" to Rorty's "mainland." In its maintenance of utopian ideals, this view agrees with Peter Koslowski that "Postmodernity postpones the final decline [*Untergang*] that is supposed to occur after the collapse of the utopian expectations contained in modernism's philosophy of history." (Hoesterey, 146)

Rorty's tolerance turns out to be a liberal intolerance towards the continued critical and radical engagement his views may suffer from utopian postmodern and feminist quarters, that is, those who still dream of revolutionizing science, i.e., life. Having dismissed emancipation for the sake of tolerance, Rorty betrays his own dominance (for which no emancipation is needed) and paranoia (pleading for tolerance so that his position of power will not be viciously attacked). Perhaps Rorty's dichotomy of tolerance and emancipation must be overcome so that an emancipatory utopia is tolerated by liberals as well as upheld by radical critics.

When feminist critics of science devise strategies through which to undermine the hegemonic position and status accorded to science they use epistemological arguments laced with political and social concerns. Alleged postmodernists such as Rorty may inadvertently undermine and derail the efforts of other postmodernists and feminist critics of science because he fails to pay attention to the need to change his attitude and orientation, to notice the psychological settings that enhance or retard possible transformations. The psychological dimension I wish to add to the political and epistemological concerns already mentioned by others pays homage to the personal commitments and convictions of individuals, their stakes in advancing critiques.

Acknowledgements

I would like to thank Cheryl Cole for her continued support and critical discussions concerning earlier drafts of this essay as well as Joseph Agassi and Debra Bergoffen whose comments helped sharpen the focus of this essay. A different version of this paper appears as part of Chapter 2 of Sassower, Raphael (1995) *Cultural Collisions: Postmodern Technoscience*, Routledge, London and New York.

References

Agassi, J. and Jarvie, I.C. (eds.), (1987), *Rationality: The Critical View*, Martinus Nijhoff, Dordrecht.

Derrida, J. (1988), *Limited, Inc.* [1977], Northwestern University Press, Evanston, Il.

Aronowitz, S. (1988), *Science as Power: Discourse and Ideology in Modern Science*, University of Minnesota Press, Minneapolis.

Feyerabend, P. (1975), *Against Method: Outline of an Anarchistic Theory of Knowledge*, Verso, London.

Foucault, M. (1970), *The Order of Things: An Archaeology of the Human Sciences,* Vintage Books, New York.

Greenberg, V. D. (1990), *Transgressive Readings: The Texts of Franz Kafka and Max Planck*, University of Michigan Press, Ann Arbor.

Guattari, F. (1986), "The Postmodern Dead End," *Flash Art* 128:40-41.

Hanson, N. R. (1958), *Patterns of Discovery: An Inquiry into the Conceptual Foundations of Science,* Cambridge University Press, Cambridge.

Haraway, D. J. (1991), *Simians, Cyborgs, and Women: The Reinvention of Nature*, Routledge, New York.

Haraway, D. J. "'Gender' for a Marxist Dictionary: The Sexual Politics of a Word" originally published 1987; "A Cyborg Manifesto: Science, Technology, and Socialist-Feminism in the Late Twentieth Century" originally published 1985; and "Situated Knowledges: The Science Question in Feminism and the Privilege of Partial Perspective" originally published 1988.

Harding, S. (1986), *The Science Question in Feminism*, Cornell University Press, Ithaca and London.

Hoesterey, I. (ed.), (1991), *Zeitgeist in Babel: The Postmodernist Controversy*, Indiana University Press, Bloomington.

Kuhn, T. S. (1970), *The Structure of Scientific Revolutions* [1962], University of Chicago Press, Chicago.

Lakatos, I. (1970), "Falsification and the Methodology of Scientific Research Programms," in *Criticism and the Growth of Knowledge*, Imre Lakatos and Alan Musgrave (eds.), Cambridge University Press, Cambridge, pp. 91-196.

Latour, B. and Woolgar, S. (1986), *Laboratory Life: The Construction of Scientific Facts* [1979], Princeton University Press, Princeton.

Laudan, L. (1990), *Science and Relativism*, University of Chicago Press, Chicago and London.

Longino, H. A. (1990), "Feminism and Philosophy of Science," *Journal of Social Philosophy*, Vol XXI, Nos 2-3:150-159.

Lyotard, J.-F. (1984), *The Postmodern Condition: A Report on Knowledge* [1979], Bennington, G. and Massumi B. (trans.) University of Minnesota Press,

Minneapolis.

Lyotard, J.-F (1988), *The Differend: Phrases in Dispute* [1983] University of Minnesota Press, Minneapolis.

Newman, C. (1985), *The Post-Modern Aura: The Act of Fiction in an Age of Inflation*, Northwestern University Press, Evanston, Il.

Ormiston, G. and Sassower, R. (1989), *Narrative Experiments: The Discursive Authority of Science and Technology*, University of Minnesota Press, Minneapolis.

Polanyi, M. (1958), *Personal Knowledge: Towards a Post-Critical Philosophy*, Harper & Row, New York.

Popper, K. R. (1959), *The Logic of Scientific Discovery* [1935], Harper & Row, New York.

Popper, K. R. (1963), *Conjectures and Refutations: The Growth of Scientific Knowledge*, Harper & Row, New York.

Radnitzky, G. (1991), "Refined Falsificationism Meets the Challenge from the Relativist Philosophy of Science," *British Journal of Philosophy of Science* 42:273-284.

Rorty, R. (1989), *Contingency, Irony, and Solidarity*, Cambridge University Press, Cambridge.

Rorty, R. (1991), *Objectivity, Relativism, and Truth: Philosophical Papers* Volume I, Cambridge University Press, Cambridge.

Sassower, R. (1991), "Postmodern Philosophy of Science: Legitimating Minority Voices," Humboldt University Summer Institute, Berlin.

Sassower, R. and Ogaz, C. P. (1991), "Philosophical Hierarchies and Lyotard's Dichotomies," *Philosophy Today*, Summer:153-160.

Toulmin, S. E. (1985), "Pluralism and Responsibility in Post-Modern Science," *Science, Technology & Human Values*, 10:28-37.

2 Against Analysis, Beyond Postmodernism

Babette E. Babich

Against analysis, beyond postmodernism: parody and pastiche

In what follows I offer a parodic brief (you'll know it by the numbered paragraphs) against analytic style philosophy just as it is that style characteristic of professional philosophy of science. I discuss the ad hoc resilience and sophisticated disdain variously operative in analytic discourse, including reviews of the maverick rhetoricism of the late Paul Feyerabend and others towards a critique of the postmodern condition in science and philosophy.

What I name continental style philosophical thinking primarily regards the historical and expressly hermeneutic style of thinking found in the reflections on science characteristic of Friedrich Nietzsche and Martin Heidegger.[1] Other continental approaches to the philosophy of science growing out of the phenomenological critiques of Edmund Husserl and Maurice Merleau-Ponty may be expected to be more congenial to analytic sensibilities as suggested by the recent resurgence of interest in the common roots of continental and analytic style philosophic thinking in Husserl and Frege.[2] An approach combining both hermeneutic and phenomenological styles with a sensitivity to the themes of mainstream or analytic philosophy of science is characteristic of the essays and books of, for one important example, Patrick A. Heelan, but also Joseph Kockelmans and Ted Kisiel, Robert Crease and Joseph Rouse, and so on – among rather not a lot of others.

Although scholars advocating continental approaches to the philosophy of science routinely refer to traditional adherents of analytic style philosophy of science, there is no reciprocal recognition on the part of analytical philosophers of science. And as a result there is no received (i.e., there is no acknowledged or recognized) tradition of continental scholarship within the professional establishment of the philosophy of science.[3] Thus the philosophy of science remains an analytic discipline, with continental perspectives excluded by the sovereign expedient of disregard, an absence of critical reference which effects the professional annihilation of scholarship. It is this factor that accounts for – that commands – the mixed style of the present essay.

Beyond such a reflection on the consequences of political domination in academic affairs, I move to challenge the coherence of analytic style philosophy as a discipline, specifically as a discipline addressed to the philosophy of science. For if analysis as a style of philosophising has a variety of faults these are most evident in connection with the real-world reference seemingly essential for a discussion of such an empirical or worldly or practical thing as science.

If I begin with an overtly modernistic parody, I conclude with a casually postmodern pastische. Casual because the conclusion challenges not the therapeutic fact but merely raises a question against the redemptive power or advantage of the postmodern condition and the disqualified master-narrative. For if marginalized discourses are to be valorized as they are in the postmodern gambit, what else is that but a pluralism of drives/discourses, each of which, as Nietzsche criticized, seeks for its part to be master? To underwrite this pluralism is to undertake the postmodern move, but one must retain sufficient suspicion to ask, with Nietzsche, whose pluralism? It is essential to ask if it can be enough to be ironic, enough to be skeptical, enough to be "open" to pluralism and to the vision of the other? What exactly happens when the last shall be first – *and knows this* –, when the margin with righteous fanfare acquires the privilege previously granted to the center?[4] Is mastery altered or is the center simply displaced to the decentered subject, the dispossessed, the excluded in all their multifarious variety and multiplicity? That is: is marginality still left, untouched, unseen, unheard? Do we, and this is the key question as we address injustice, as Horkheimer and Adorno warned, simply invent other Jews as we decry anti-Semitism? Is our denunciation our own and only reward? Why, if difference is to be celebrated, do all differences wind up each on the same level, gender, race, class? Are women the same as (have they the same needs, the same oppressions, are they a class the same throughout, a class at all any more than) blacks, are blacks the same as Jews, are gays the same as lesbians? If they are or can be or can be imagined identical, merely similar, or even said to be "in the 'same' boat," it is because of a polarization from a central reference, a referentiality unquestioned by every question raised against it. For if the other can be thought at all, if the other can be thought as Other, it is only because the centrality of the master signifier, the white, the anglo-saxon, Euro-ideal male, scholar and bourgeois representative, is not and ultimately can never be in question. *As alternative, as the other excluded, the center and the margin remain an issue to be decided one on one in an ordinal binary opposition.* And this is exactly not pluralism, where pluralism can only be thought in irretrievably perspectival terms, in all its complexity, where the truth of pluralism is exactly non-simplistic in its multiplicity and truth (in Nietzsche's terms) a lie in a lying dialectic of lies.

Thus I argue that the postmodern project reports the condition of knowledge, just as implied by the commissioned subtitle of Jean-François Lyotard's *Postmodern Condition.* It is not however a solution, if only because the thought of resolution is a singularly analytic, which is also to say, absolutely modern ideal.

22 paragraphs against analysis: A stylistic propædeutic and prospective

1. The project of analytic style philosophy, whether the analytic frame be that of ordinary language or logic, is clarity. By clarity is meant clarity of expression. For Wittgenstein who coined the effective *Leitmotif* of analytic style philosophy in his *Tractatus*, "everything that can be put into words can be put clearly." (Wittgenstein, 4.116) Thus, philosophy, "the critique of language," (4.112) is "the logical clarification of thoughts." This clarity may be attained by definition (or fiat), but a clearly expressed proposition is, even if a statement of a problem, surely less mysterious than an unclear statement of the same perplexity. And just as the name analysis suggests, the point is to reduce or dissolve philosophical problems.

1.1. Beyond an idealized articulative clarity, analytic style philosophy enjoys the streamlining images of two additional regulative ideals: inter-subjectivity and verification. Intersubjectivity eliminates mysticism, esotericism, private languages and inaugurates (as a solipsism writ as it were upon the world) the analytic problem of "other minds." And by the simple expedient of bringing the "charwoman" or the "man in the street" – however quaint, however rhetorical in intent and practice – into the hallowed circle of Robert Boyle's gentlemen observers and the noble assurance of objectivity, the intersubjective emphasis leads not to a circularity among elite subjects, but ordinary language philosophy instead.

1.2. For the second regulative ideal, as the question of the intersection between word and object, verification is an epistemological issue, an ontological question, and for analysts, a metaphysical quagmire. The statement, "The meaning of a proposition is its method of verification" leads in its Tarskian formation to nothing else again but the ideal of clarity. With a thus impoverished empirical ideal of presumedly unproblematic reference (observation "sentences") there are propositional objects in the world of the analyst but only patterns or atoms of experience: pink patches – or pink ice-cubes, a once-outré Sellarsism – or gruesome impressions.

2.   The analytic ideal of the clarification of meaning is not only or ultimately a matter of the clarification of terms. Rather what is wanted is the reduction of problems, their revelation as pseudo- or as non-problems. All problems that cannot be clearly stated are problematic statements.[5] Hence all problems that can be counted as such are analytic and hence lysible.

3.   The success of analytic philosophy is intrinsically destructive. By definition: the philosophic project itself is repudiated in its ambitions, reduced to trivialities, and thereby overcome. This is why Wittgenstein's ideal involves disposing of the ladder (of analytic method) after reaching the heights of clarity.

4.   By success is meant nothing more than the application or employment of analytic philosophy in practice.

5.   This is not true of all philosophic ventures (despite the Hegelian inclination to assert the contrary). Hence the success of the Heideggerian project of the destruction of metaphysics does not equal or reduce to the destruction of Heidegger's project. Nor indeed does the success of the more notorious and more likely instance of deconstruction conduce to its own end. To the contrary.

6.   At issue in the analytic project is the end of philosophy – taken in decidedly non-structuralist guise. For analytic philosophy: all of metaphysics, together with the traditional problems of philosophy, is, as an accomplished and desired deed (*philosophia perennis confunditur*), already at an end and by definition (as meaningless or non-verifiable). What remains or is left over is to be resolved by analysis. Since traditional philosophy is set aside along with its perennial questions – these are philosophical questions disqualified as such because of their resistance to analysis/resolution – an end is also made of the tradition of philosophy. In the place of the tradition we find science. Science, for its part, is an empirical enterprise, but devoted to clarity and committed to intersubjectivity (coherence or making sense) and the logical problem of verification appears to be the principle or fundamental concern of logical analysis or (analytic) philosophy of science. Hence the received view in the philosophy of science is developed in the analysis of theories in the hypothetico-deductive programme.[6]

7.   Science is a suitable subject for analysis proximally because it is itself a body (theoretically expressed) of clearly stated propositions or claims that describe for language users (intersubjectivity), the structure of the world and are either true or false in that connection (verifiability). Science itself, it is said, is empirical analysis, a prime example of the productivity of analysis. Circularities would seem to abound here, as cannot be helped

when tautology is one's stock in trade, but if they are not affirmed as they are in hermeneutic "circles," they nonetheless provide the advantage of certainty. As Philipp Frank, one of the founding members of the Vienna Circle expressed the former virtue of scientific analyticity, in a statement combining the insights of Mach with the Kantian conventions of Duhem, "the principles of pure science, of which the most important is the law of causality, are certain because they are only disguised definitions."[7]

8. Empirical observation and experiment together with logical analysis is canonically held to decide the value of a claim or theory. Thus analytic philosophy of science has essentially been conducted within the spirit of the Vienna Circle. Despite Mach's "physicalism" the members of the Vienna Circle, in the words of one commentary, "wrote as though they believed science to be essentially a linguistic phenomenon."[8] This predilection for "language" be it ordinary or logical, together with a naïve view of direct observation (i.e., observation sentences) means that the analytic concern of the philosophy of science has been restricted to the analysis of theory, in a word the received view or hypothetico-deductive nomological ideal of science (theory).

8.1. Analytic statements are by definition tautologous and assert nothing about the world. This is their virtue and at the same time, this is their impotence. Empirical statements are what is wanted in science.

9. This focus on the elements of language – not Machian physical-physiological elements – dramatizes a rupture between language and world (the limits of language) which as the essence of tautology or logical linguistic self-reference is not problematic when what is analysed is language use, the game or its rules, but only when what is analysed are empirical matters.

10. The socio-historical turn in the philosophy of science, identified with, among others the otherwise analytically sensitive Hanson, Kuhn, and Feyerabend together with (and this is what must be seen to be decisive) the so-called strong programme of the sociology of science (not knowledge) has yet to be accommodated in the philosophy of science. It is this that constitutes its continuing crisis. This crisis corresponds to its philosophical failure, a philosophical failure tied to the fundamental schizophrenia of its analytic origins. Despite a fascination with language, and thereby, in a kind of return of scholastic nominalism, with certainty and the idea of eliminating philosophical problems by the expedient of linguistic or logical clarification, a positive empirical reference remains relevant to science. This reference to empirical matters in the relevance of scientific practice is what analytic philosophers of science mean by naturalism.

11. Naturalism, which for Tom Sorell (1991) is itself a form of scientism, is

not philosophically distinguishable from the normative or analytic issues of verification or legitimation. The ultimate reference of the philosophy of science remains "natural" or actual science. As Rom Harré observes, as disingenuously and as plainly as any analyst could wish, "the philosophy of science must be related to what scientists actually do, and how they actually think."[9] The imperative to express such a relation to actual scientific practice derives not from ascendent realism but rather from the socio-historical turn that comes after the linguistic turn.

12. The socio-historical turn seems unrelated to the analytic or linguistic turn. Yet the conviction held by philosophers of science from Carnap to Hempel to Suppe and beyond, that science is a formal, logical, or linguistic affair was not the result of a devotion to logic as such. Empiricism or positivism as it was understood by Auguste Comte – the first "positivist" – embraced a positive reference to facts. Thus Hacking recalls Comte's 'positivity' as "ways to have a positive truth value, to be up for grabs as true or false." (Hacking, 12) The ultimate appeal of Wittgenstein's logical programme of linguistic therapy (analytic clarity), combined with Mach's physical critical-empiricism for the members of the Vienna Circle was in the celebration of and application to practical, actual science. Only in the era of the triumph of scientific reason would such an analytic programme work as successfully and despite patent internal contradictions as long as it has without drawing undue attention to those same contradictions.

13. For even if the project of analytic philosophy had been shown to be bankrupt from a realist or empiricist or naturalist point of view, as long as science is associated with reason, and reason or rationality is equivalent to logical analysis, it will be analytic style which gives the imprimatur to proper philosophical approaches to the philosophy of science, no matter the actual success of analysis in offering an account or philosophy of science. For this reason Rudolf Haller points out, talk of verification – an analytic specialty – works as a Popperian "*aqua fortis* for separating good and bad talk in science and philosophy." (Haller, 266) Analytic talk remains the dominant strategy of legitimacy and distinction in the demand for clarity and coherence. And it is fundamentally flawed not just for the tastes of those who are not convinced of the salutary or edifying values of clarity and coherence but according to its own rationalistic terms as well. For there is no obvious connection between deductive (or inductive or abductive) logic (or grammar or language) and the world. Assuming without the metaphysical faith of a Mach or the teasing leap of a Feyerabend such an elemental or obvious connection as axiomatic or given, the analyst ends up so preoccupied with refining his or her logical tools, that he or she forgets having renounced contact with the world.

36

14. The history of scientific theory and experiment, popularly known as the "scientific revolution" is not the project of pure theory or metaphysical speculation. Instead, it is physical or "'physicalist." It is the history of factual observation (controlled experiment) and theoretical explanation. For analysts, the former are to be expressed as empirical statements and with the verification of such observations, converted into so-called protocol statements to which experimental or theoretical conclusions reduce now as theory with full-fledged (so analytic) propositional content. This is the ideal analytic recipe that guarantees scientific control (progress). This same programme frees humanity from its (self- or deity-imposed) bonds of superstition and inhibition.

15. Yet it is just as clear from the reference to observation and experience that the history of experiment is also the history of power, manipulation, illusion. The project of experimental progress is in short that of the history of technology.

16. Separating the theoretical ideal of Newton's *hypothesis non fingo* from Boyle's celebration of neutral and observationally-objective (subjectively-independent or intersubjective) experiment is the tacit and practical rôle of evidence. This introduces the realist question of what evidence? evincing what? and the naturalist's but still more relevant sociologist's question of evidence obtained by and for whom? The issue of evidence is to be contrasted with theoretical truth. The last remains a matter of configured, what Nietzsche would name *fingirte*, hypotheses.

17. More than a conceptual net, one has an array of hypotheses and praxes, so that the infamous impotence of the *experimentum crucis* to decisively refute a scientific hypothesis or theory blinds one to the already given and far more pernicious matter of focal, selective choice. A given conceptual net is woven out of if not whatever we please surely what we happen to have on hand. Moreover, there is no way to imagine, beyond Duhem-Quine, as Davidson points out in his essay "On the Very Idea of a Conceptual Scheme," that this or any other conceptual scheme represents the way things are (or are not).[10] What once represented a psychological strategy, (proto-Piercian) quiescence of belief, *ataraxia*, or calming, Stoic equipollence, is today a feature of crisis. What works as therapy in one context is, as the ancient Greeks knew perhaps best of all, death in another.

18. More devastating than Duhem's instrumental critique of the use of experiment is that which follows from Mach's *empiriokriticismus* and in his view – a perspective shared by Polanyi, Hanson, and Fleck, and historically articulated by Kuhn – the ideal of a quasi-artistic invocation of research style and experimental tactic or technique or knack (also to be heard in Mach's conviction that experimental practice could not be taught

– just as artistic talent is not communicated by instruction) in the life of the researcher. The notion of scientific schools, "invisible colleges," *Denkkollektiven*, knowledge communities, and so on, offer particular inspiration for sociological studies and observations. The question of what, in Harré's words, "scientists actually do" remains in a scientific era the ultimate issue. It is this and the tracking of the question as a matter of a research discipline – not among philosophers, analytically or otherwise inclined but scientists, albeit scientists of a social kind pursuing a discipline focussed upon scientists themselves, – which may be said to have added a kind of last straw to the woes of analytic philosophy.

19.   Ultimately, the method of analysis is philosophically and scientifically impotent. Analysis has as it goes along, and this by its own rights, "less and less of what to analyze."(Bar-On, 1990, 260) Note that reduction as such (the disgregational, dissolving, when not always dissolute gesture implied in the idea of analysis) was not opposed by Mach who was with Richard Avenarius an enthusiast of the ideal of a scheme he imagined reflected in nature itself. But in spite of this latter realist (and here: metaphysical) resonance, Mach's ideal of *Denkökonomie* preserved its methodic function: it was a tactical, heuristic ideal, not an analytic end that simply reduced a problem to its linguistic, logical components and left it at that as if solved, whereupon one could, as it were, throw away the ladder. For Mach, everything could be reduced if one could assume as he did and the Vienna Circle did not, that everything was convertibly elemental. The unified scheme of the received view of the philosophy of science reflected not Machian elements – constituting the physical, physiological, psychological world – but observation sentences linked by correspondence rules to theorems, beginning and ending with units of logic/language. The world here is what is symbolizable, coordinative, re-symbolizable; neither fact (*Tatsachen*) in the end (linked as these are with theory) nor thing (whatever a thing may be).

Against analytic philosophy: disclaimers

After such a parodic feast, it seems plain and only fair to add that analytic philosophy, apart from the philosophy of science in any event, is much further evolved these days than once it was. One no longer spends the whole of one's analytic philosophic energies analysing (according to the exactitude and focus that is an irreducible part of such methodic precision) statements such as "The cat is on the mat," but one allows oneself the still unexhausted fit of fantasy indulged in by Tom Nagel who wondered "What Is It Like to Be a Bat?" (with

38

its predictable if not quite logical sequel: "The View From Nowhere"). Or, more appositely, one might follow David Lewis who very charmingly begins his "Attitudes 'De Dicto' and 'De Se'" with the observation against expectations – that is, ladies and gentlemen, just to be sure that you do not miss it – a joke, a piece of wit: "If I hear the patter of little feet around the house, I expect Bruce. What I expect is a cat, a particular cat." (Lewis,133) Of course, to the point of punning, the patter of little feet, not to mention the talk of expectations, refers, for speakers of ordinary idiomatic English, to children. The joke brings in Bruce, the cat, and the reference to the cat takes us to the mat and the matter of reference. Lewis's observations are about Meinongian attitudes which is to say or to be read as shorthand identification for psychologism (a bad thing) or intentionality (possibly a good thing, provided the intended intentionality is not that of the late Husserl but rather the early, now redeemed as the Frege-like, and almost analytic Husserl). In this case the attitudes are explained as incomplete where such expectations may be diversely filled in divers houses (Lewis's specialty is possible worlds so an array of possible houses is no strain for him). These attitudes then are best rendered, so Lewis, as having "propositional objects." We recall that for analysts, propositions are technical devices, having, as sentences do not always have, logical objects.

Note the utility of the style of this kind of talk for analytic purposes. It is because we may be expected to be concerned with whether and what we mean with what we say (the allure of this concern is not least won from precisely that clean or neat reference and conceptual (if none too taxing) ideal of analytic clarity, which in turn consists in the play between notions of the expected and what is as such, in other words and in another sense, *de dicto* and *de se*) without at the same time and in fact actually meaning anything in particular by what we are saying. Thus we talk about cats, bats, and brains in vats. The result of this linguistic explosion of deliberately irrelevant reference permits us for the first time if also and admittedly only for the nonce to consider meaning as such.

All of this can make for very entertaining reading (especially when it is David Lewis one is reading) but this appeal does not go very far – and this returns us once again to the problem at hand – when what is at stake matters as much as science does. It is then that the analytic style, tactic and schematic, runs into the proverbial ground and it does so without necessarily drawing attention to this fact among practitioners of the philosophy of science.

The idea of "going to ground" or "to seed" or "to hell in a handbasket" or better, with reference to analysis, of disgregation – whereby the practice of analysis ends up with "less and less of what to analyze" – is manifest in the whimpering perpetuation of things as usual. This is the way the world ends in the face of everything: a kind of heat death which Nietzsche, a famously non-analytic philosopher, called nihilism.

Still, and yet, following the historical or interpretive (but not and the difference between terms is significant: hermeneutic) or sociological turns, it seems that no practitioner of the formerly recieved view in the philosophy of science can be found on the books. The problem is (in the parlance of informal fallacies) a straw man. Analysis, it would seem, has long since been overcome. Against analysis? Against method? Who – we might ask ourselves – isn't?

Indeed, some time ago a mainstream collection appeared with the title *Post-Analytic Philosophy*. Contributors (and putative post-analysts) included Putnam, Rorty, Nagel, Davidson, Kuhn, et al., all of whom were and still are said to have – and were accordingly lionized for their intellectual integrity for having done so – abjured analytic philosophy (and all its works). Yet it is evident enough, where what matters on the terms of analysis itself is style, *analytic* style and precisely not – such is the formal ideal – substance or content, that no one of the above is, in fact, anywhere near post-analytic.[11] It is important to note that one can persevere in one's allegiance to the analytic ideal and remain an analyst without the analytic program – an essential survival strategy when its traditional adherents (Putnam, Nagel, Davidson) concede the flaws of the program.

Such a righteous confidence is characteristic of established power elites and a typical retort ("argument") to a critique such as the foregoing need do no more than dispute the given definition. Thus one notes: X is averred (analysis is X). But, one avers to the contrary, analytic philosophy is in fact -X. Thus one maintains that analytic philosophy (here: -X) is actually some other thing than had been claimed (and -X is pretty broad), perhaps χ, or some other thing.

These are analytic tactics: they sidestep the question, shifting debate to formal (analytic) grounds and they do so in perfectly good conscience (albeit perhaps not in perfect good faith). Denying the relevance of critical charges only sidesteps the question. Like talk of full-fledged postmodern science or pushing it further, postmodern philosophy of science, conventional talk of the end of philosophy, especially the end of analytic philosophy, is a plain exaggeration. For even if, politically and otherwise, these are lively times we live in at the end of this century: if we are ideologically bound, by at least popular convention, to be pluralistic, open to new ideas, different perspectives on east-west, other ideologies, and if we are therefore, whether-we-like-it-or-not, living in a "postmodern" world, be it noted that neither Richard Rorty nor Jacques Derrida nor the unnamed demon of irrationalism, or relativism, have genuine influence in the prime analytic domain of the philosophy of science. Nor are specialists in "irrationalism" (read: continental-aka-hermeneutic-style philosophy)[12] hired at the university level for whatever few positions there are in philosophy. The majority of university departments remain analytic and when they hire, even when they hire for positions specified as dealing with more or less continental thinkers (e.g., Husserl, perhaps Schutz, or Merleau-Ponty, hardly ever

40

Heidegger, never exactly Nietzsche), hire retread analysts (UK phenomenologists or German trained analysts – the last even more fun than the former.) And if (again: of all philosophical subdisciplines) the philosophy of science is not non-analytic, neither can it be said that the philosophy of science is postmodern (either "already" or *in nuce*).

Against analytic philosophy of science: Say! Didn't Feyerabend do this one?

The philosophy of science is soberly modern, by definition and design. In this way, too, discontent with the modern, or the cry for alternative approaches is rarely sounded in a discipline so literalistic that most of its practitioners are convinced that Paul Feyerabend (or even Thomas Kuhn) is a mortal enemy. This literalism (an exact tactical advantage among analysts) means that Feyerabend's book, which he did, after all, title *Against Method*, is taken as a treatise composed to methodically argue, as treatises argue, against method. Literally. And thus, metonymically (by sheer associative thinking), against science.

Where scientific knowledge is knowledge methodically obtained, an anarchistic project is ranged against scientific advance. But Feyerabend was actually, as he happily detailed again and again, arguing not in the manner of a treatise or for his own views but rather against alternate views or, later, the views of his critics or opponents.[13] Feyerabend's own views are (rather typically in this rhetorical context) elided by repudiation.[14] Feyerabend was arguing for a Machian physicist's or scientific practioner's concession to the aesthetic innovation actually characteristic of the scientist's application of method. *Against Method* was thus written for the sake of scientific invention, and although Feyerabend has no kind words for Rom Harré, as a man,[15] he does share many of his ideas, especially Harré's dedication to the importance of giving a philosophic expression of "what scientists actually do." In Mach's case (Ernst Mach, we remember, was Feyerabend's all-time favourite non-philosopher; readers with other affectations may prefer to invoke Michael Polanyi's tacit dimension or else Ludwik Fleck's 'style'), this attention to practice involves adverting to that which "cannot be taught."[16] Even if Feyerabend's thought is astonishingly global enough and where Feyerabend himself was certainly generous enough to regard science as only one of many human endeavors, and even to argue against its primacy in today's societies, Western and other, Feyerabend was never for all his cosmopolitanism ("cultivation" or magnanimity) anything like an enemy of the ideals of the philosophy of science in any sense.[17]

"Anything goes" is the call to arms promoting nothing other than the ideal and project of science. Progress is not thereby devalued where method is denied in a canonic sense. Instead, given the advantages in practice of the ad hoc for jazz players and for experimental and theoretical scientists, the anarchic ideal in conjunction with efficacy is offered in the spirit and for the sake of progress. The world of art, of painting and theatre, so Feyerabend is convinced, is already complicit (if art is not "free-style" what is?) and Feyerabend invites the other realms of culture, science included, no better and no worse, to come along in the same spirit. Lyotard – and I will need to return to this point below – reasons similarly to find science, indeed information science (the exact cybernetics of Heidegger's longest nightmare)[18] playfully, redemptively postmodern.

Feyerabend may be listed among analytic philosophers, for whom the logical (or linguistic including ordinary language) clarification of philosophic problems has special application to science. The ideal is to understand scientific progress, attributable, it is thought, to the special use of the scientific method.[19] That this method would seem for Feyerabend (and Norwood Russell Hanson and even, on the continental side of things, Patrick A. Heelan) to be a matter of improvisation is one of those complicating, confusing details anathema to the ideal of analysis and clear expression. It is not irrelevant that Feyerabend and Hanson and Heelan began their intellectual lives with studies in mathematics and physics without however as so many other physicists-cum-philosophers of science being lulled by the idols of the tribe. Note that Feyerabend's cutting or avant gardist edge against reactive analytic thinking derives from a praxical affinity to science as an experimental and hence provisional enterprise.

Beyond analysis: towards a postmodern critique

I claim that analytic philosophy fails as the ideal of logical positivism due to the dogmas of empiricism (however many), in particular, that it fails as analysis not by default but intrinsically as a circular, self-consuming philosophic style. Nonetheless, even so, I also maintain that the tradition remains the Anglo-American (i.e., *the*) style of doing philosophy. Just what it is that one is doing: analysing logic, ordinary language, or what have you is up for grabs. The question here might be, what has all this got to do with the philosophy of science? For if it is assured that science "works" – as a research enterprise – cannot it not also be said that the philosophy of science works?

The problem is that the philosophy of science does not work – not, that is, without qualification. Not if its object is to explain or to understand the success of science. And not if its object is to analyse the goals and methods of the sciences or if the philosophy of science is taken, once again recalling Harré's modest, naturalistic formulation, to refer to what "scientists actually do." For the practical success of the natural sciences is not its use of the (so-called)

42

"scientific" method alone nor even its mathematization or formulization, although publication of the latter and assertions of the former enhance academic and political prestige.[20] The recent explosion of studies under the rubric of the sociology of knowledge/science, indicates that the progress and practice of science is rather like that of any other social activity. Science, if regarded as a privileged (putatively unique) cultural practice rather than a singular, singularising trans-cultural truth project, becomes a social, that is, all-too-human affair. Thus the issue is not simply epistemological or methodological but complex, practical and anthropological. As Bruno Latour expresses this with his usual anthropologist's (and very scientific) disciplinary hyperbole: "Suddenly, we look at our sciences, our technologies, our societies, and they are on a par with what anthropology has taught us of other cultures." (Latour, 288)

Classic continental approaches to the philosophy of science echo many postmodern critiques, where both approaches to science challenge the claim of science's very singular prerogative. The modern canon for legitimating discourse – that is, again: rationality, objectivity, truth, progress, the scientific schematism of limit and hierarchy – is called to account by both broadly continental and specifically postmodern critiques. Apart from an appeal to the unquestioned value of its own authority, science's response to such challenges is perforce limited: the very scientificity of science, the value (or values) of science, is called into question here. The neutrality (as well as the limits or even the ultimate "terminability") of experiment is questioned. The objectivity of the experimenter\observer is challenged. The morality of the project and procedure of science as such is scrutinized. In all, the hegemony of mathematical and physical science as the single best way of knowing is criticized and so undermined (as erstwhile axiom). Commitment to even the possibility of postmodern critique appears to end the undisputed authority of science.

Whether it also and necessarily means the end of science or its marginalization (i.e., dispersed within an array of pluralist economies of knowing) is not equally clear. For Lyotard (just as for Stephen Toulmin), where it is argued that science itself has taken the postmodern turn, science (and rational discourse) becomes itself a representative expression of the postmodern condition.[21] In this way, contemporary "post-modern" science continues as it always has,[22] i.e., as more or less modern (as post), recognizing, as Toulmin has it, that there is no pure starting point or "scratch line," and conceding – or affirming – as both Nietzsche and Wittgenstein have argued, the inevitability of certain ambiguities as a bar to ultimate logical transparency or clarity. The often touted transformation of the physical and information sciences under a newly minted concern for "chaos theory" and so-called fuzzy logic reveals a natural science attuned to integrational holistic visions, gender sensitivity, and environmental friendliness, in place of the the old modern scientific aims of explanation, prediction, and control. Yet we do well to avoid the glib security of those theorists of the postmodern who claim the redemptive power of these accomodations. It should be enough (but is not, such is the power of an associative connection) to note that the affirmation of fuzzy logic (or "chaos" theory) fuzzes or obscures the key conservative and hence predictive element in logic, reflecting the requisite conditions for measure and successful calculation rather than any departure from the modern Cartesian ethos of certainty in the service of control. Toulmin's reprise of Montaigne's sophisticated humanism in his *Cosmopolis* may be little more than a softer Cartesianism, a scientific humanism with all enlightenment ideals intact, refurbished for the new millennium. (Toulmin, 1990)

Like Toulmin and Lyotard, Habermas and other critics argue that continental philosophical thought, including related postmodern critique is nothing but a further development of the modern philosophical tradition of critique and throughly modern demystification. As such, so-called postmodern critique is only an extension of and no break from the modern enlightenment project of demystification. Yet if Habermas and Co., including Richard Rorty, be right in this, they must find themselves willy-nilly aligned with postmodern authors, from Lyotard and Baudrillard to Toulmin and a host of others. In this description, postmodernism appears more friend than foe to the sciences and consequently, by extension, to the philosophy of science. Insisting that we question the traditions of science and exploring the ways in which science has become dogmatically authoritarian, postmodernism as a super-enlightenment, meta-meta-discourse thus poised against itself (as a claim concerning the impossibility of metadiscourse), can recall science to its original motives. In such a mega-modern fashion, postmodern critique might then be seen to be posed less as an attack on science (anti-science) than as an exposée of certain naïvetées.[23]

Yet exactly here it may be incumbent upon today's critic to suggest that a certain irony, a very postmodern incredulity be reserved for proposals concerning the postmodern status of knowledge, particularly science, certainly information science. For, apart from Lyotard's (or Baudrillard's or McLuhan's or any other media expert's) ecstatic enthusiasm for the liberating virtues of the information revolution, the idea itself is patently overblown.

Virtual reality by another name is the simulacrum. The thing about the simulacrum (a computer game, surround sound, multi-media computer graphics) is that it is very manifestly a substitute, like driving a play automobile at a video arcade.

The computer image is coded – read and interpreted with perfectly hermeneutic alacrity – as it is in every other sphere of "real world" perception, but coded *as* unreal, *as* an image. It takes away not at all from the realistic charge (or kick) of such virtual images that they are palpably inferior (impalpable) substitutes. For the kick is exactly that they be as good as they are. "Surround sound" sounds as if one *might be* in a live concert. To sound this way, of course, given the accoutrements of the ordinary living room, drapes, couches, carpets, and given the distractions of a picture window or a nearby kitchen conversation, it has to be, *and it is*, larger than life. It is in this overwhelming imaging that the realism of the substitute consists.

The interactive CD-ROM game, *Myst* is currently touted as the most imaginative and best "on the market."[24] What makes *Myst* best (at least at the time of this writing) is firstly its imaginative conceit but more critically, as this is what makes the concept work as such, it is its density of graphics images, their sheer number, as stored and playable on CD-ROM. Thus the game plays – when played on computer monitors with the best video accelerators and graphics cards possible – like a film or better like an interactive computer graphics cartoon, which is what it is. The world of *Myst* is not in fact realistic (where "realism" in life and in artwork, please note, with a nod to Benjamin, is exactly photographic or filmic two-dimensional realism) but the bloated surrealistic, pixel-determined world of computer graphics games. From the opening scene of a male silhouette's plunge into a cartoonish rift of stars, to the landscapes and the interior and exterior architecture, *Myst* is comprised of computer graphic images of the cartoon-unreal caricaturing a sci-fantasy, deserted island retreat. It may seem surprising, though it ought not be, that the talismanic image is the book, and the game's high point-and-click achievement is that of taking a book down from the shelf, of turning its pages, one mouse movement at a time.

The simulation (and hence the improved utility) of the book is the goal of CD-ROM as a medium. This goal is manifest enough in the complete CD-ROM research editions of the philosophic oeuvre of, say, Hobbes, just as it is manifest in a CD-ROM telephone book. What the CD-ROM version permits is not reading as such but the dispensation from the necessity of the same, finding given words in a global search or sweep.

The issue here does not concern the message but the medium and the consequences of an automatic credulity, better a belief not in metanarrative as such but in megabytes and still and yet in redemption through technology, ever more rarified to the internet, to email, to spreadsheets, to wordprocessing and the labile and virtual text.

Lyotard, Habermas, Rorty, Taylor and so many others tell us in very different ways that the current information age is the age of liberation. Liberation, for enlightenment thinking, is exactly progress. But the point is to press a question against our credulity in automation, as a credulity in the electronic order.[25]

What then? Then nothing. Like science, the modern project of the philosophy of science is willy nilly, like it or not, become a postmodern gambit.

This is both more and less than Lyotard might have imagined. The postmodern is not a resolution of the modern, it is its current and just as paradoxically as Lyotard had imagined, its ennabling *condition*.

The task for the next millenium, at least at its start, is addressed to the failure of the modern as an imaginary fault; postmodern, if at all, by default. For we are hardly postmodern already – despite the assertions of Lyotard, Venturi, Eco, Jencks, and even the present author. We are not free from our once and former tutelage to the myth of the modern, the metanarrative of progress, of reason, and of science. At best unwittingly free from these same metanarratives, the liberation we have is that of anomie following distraction and disappointment. Human maturity that is not nihilism, that will mean rediscovering, as Nietzsche observed, "the seriousness one had as a child at play."[26] Because of the extreme mastery and exceeding delicacy involved in play (consider music but also any performance art, or sport but also chess or other competitive games), the challenge of the postmodern is an absurdly, impossibly innocent child's play.[27]

References

Babich, B. E. (1995), "Heidegger's Philosophy of Science: Calculation, Thought, and *Gelassenheit,"* in Babich (ed.), *Heidegger from Phenomenology to Thought, Errancy, and Desire*, Kluwer, Dordrecht.

Babich, B.E. (1994), *Nietzsche's Philosophy of Science: Reflecting Science on the Grounds of Art and Life*, State University of New York Press, Albany.

Babich, B. E. (1994a), "Philosophy of Science and the Politics of Style: Beyond Making Sense," *New Political Science*, 30/31:99-114.

Babich B.E. (1993), "Continental Philosophy of Science: Mach, Duhem, and Bachelard" in Kearney, R. (ed.), *Routledge History of Philosophy: Volume VIII*, Routledge, London, pp. 175-221.

Bar-On, A. Z. (1990), "Wittgenstein and Post-Analytic Philosophy" in Leinfellner, Haller, et al. (eds.) (1990) *Wittgenstein.Eine Neuebewertung: Towards a New Revaluation* II, Vienna, p. 260.

Davidson, D. (1980), *Reference Truth and Reality: Essays on the Philosophy of Langauge,* Routledge Kegan Paul, London.

Feyerabend, P. (1989), *Farewell to Reason*, Verso. London.

Feyerabend, P. (1981), *Erkenntnis für freie Menschen*, Suhrkamp, Frankfurt a.M. English translation published (1987), *Science in a Free Society*, Verso, London.

Heelan, P. A. (1983), *Space-Perception and the Philosophy of Science*, University of California Press, Berkeley.

Hacking, I. (1992), "'Style' for Historians and Philosophers," *Studies in the History and Philosophy of Science.* 23/1:1-20.

Haller, R., Schurz, G., & Dorn, G. (eds.), (1991), *Advances in Scientific Philosphy*, Rodopi, Amsterdam.

Harré, R. (1972), *The Philosophies of Science*, Oxford University Press, Oxford.

Latour, B. (1992), "One More Turn after the Social Turn . . ." in McMullin, E. (ed.), *The Social Dimension of Science*. University of Notre Dame Press, Notre Dame.

Lewis, D. (1979), "Attitudes 'De Dicto' and 'De Se'," in *The Philosophical Review* 1979 No. 9. Also in (1983), *Philosophical Papers: Vol. 1*, Oxford University Press, Oxford.

McGuinness, B. F. (1989), "Ernst Mach and His Influence on Austrian Thinkers," in Gombocz, W., et al. (eds.), *Traditionen und Perspektiven der analytischen Philosophie*, Hölder-Pichler-Tempsley,Vienna, pp. 149-156.

Nietzsche, F. (1980), *Sämtliche Werke. Kritische Studienausgabe,.* Vol. 1-15, de Gruyter, Berlin.

Maia Neto, J. R. (1991), "Feyerabend's Scepticism," *Studies in History and Philosophy of Science*, 22/4:543-555.

Redner, H. (1987), *The Ends of Science*, Westview Press, Boulder, Co.

Richardson, W. J. (1968), "Heidegger's Critique of Science," *The New Scholasticism*, LXII:511-536.

Rouse, J. (1991), "Philosophy of Science and the Persistent Narratives of Modernity," *Studies in History and Philosophy of Science.* 22/1:141-162. 1991.

Sorell, T. (1991), *Scientism. Philosophy and the Infatuation with Science,* Routledge, London.

Suppe, F. (ed.), (1974), *The Structure of Scientific Theories*, University of Illinois Press, Urbana.

Toulmin, S. (1990), *Cosmopolis: The Hidden Agenda of Modernity*, University of Chicago Press, Chicago.

Toulmin, S. (1974), "The Structure of Scientific Theories" in Suppe (1974).

Wittgenstein, L. (1974), *Tractatus-Logico-Philosophicus*, Pears, D.F. & McGuiness, B. F. (trans.), Routledge & Kegan Paul, London.

Notes

1.   This essay is related to a series of essays which necessitates either redundancy or cross-reference. As the lesser evil, I have sought here and below to minimize the former. For a discussion of Nietzsche, see Babich (1994); for Heidegger, see Babich (1995).

2.   See references, Babich (1993). Citing a recent doctoral dissertation by Brian Mattingly, Patrick A. Heelan takes Mattingly's arguments to imply

that continental style thinking includes the hermeneutic language phenomenology of the later Wittgenstein.

3. Because none of the above mentioned names from Nietzsche and Heidegger to Husserl, Heelan and beyond are featured in discussions of or even included in bibliographies of the philosophy of science proper, it is as if continental style approaches to the philosophy of science did not actually exist. On this see, Babich (1994a) and (1993).

4. This is the problem with many postmodern, pluralist, feminist moves, especially those made in the tradition of Foucault. On the advantages of the feminist postmodern critique, see Raphael Sassower's excellent discussion of Donna Haraway above (pp. 25 ff.). For a discussion of the limits of such automatic writing-in, as it were, of the disenfranchised other or position, see Babich (1994a).

5. To vary David Lewis' expression in his "Attitudes 'De Dicto' and 'De Se'" of the implications of Wittgenstein's notion of expression and clarity: if it is possible to have unclarifiable (unanalytic) problems but no unanalysable propositions, anything propositionally articulated – which in this sense means clearly expressed – can be analysed. As David Lewis states the virtues of propositional knowledge, "...if it is possible to lack knowledge and not to lack any propositional knowledge, then the lacked knowledge must not be propositional." (*Philosophical Papers: Vol. 1*, p. 139).

6. The received view has had an exceedingly short tenure for a defining philosophical structure: the Cartesian account of the role of the pineal gland could claim both a lengthier reign and greater fecundity. See Frederick Suppe's "The Search for Philosophic Understanding of Scientific Theories," in Suppe, pp. 3-232.

7. Frank, "*Kausalgesetz und Erfahrung*," *Annalen der Naturphilosophie*, 6:443-450.

8. Dilworth, Craig "Empiricism vs. Realism: High Points in the Debate During the Past 150 Years." *Studies in the History and Philosophy of Science*. 21/3:431-462. p. 224.

9. Harré, p. 29. Emphasis added.

10. Davidson, "On the Very Idea of a Conceptual Scheme," pp. 183-98.

11. Analytic style as such refers to little more than the ideal of expressive clarity.

12. This is especially true of that kind of *continental style philosophy* associated not with the softer theories of ethics or the political world (critical theory and so on) but with analytic turf-encroaching topics such as epistemology, in Husserlian phenomenology and, via Nietzsche and Heidegger, hermeneutics.

13. Thus when Ernst Gellner complains that Feyerabend "cannot lose," he is right. See Feyerabend, 1978, p. 142. Feyerabend does to Gellner what he does to all his critics, he holds them to critical standards from which he exempts himself. But it is his critical "turn" when he does this.

14. To wit, Feyerabend's repeated (analytic style) declaration: "these are not my views..." This is an assertion one should know better than to take at face value even if Feyerabend's commentators are still assiduously engaged in word frequency counts of dadaism and anarchism and all references to voodoo and astrology in Feyerabend's earlier and later works) See, for example, Maia Neto, p. 544.

15. Thus Feyerabend denounces Harré's perturbation concerning various "asides about women, about friends and colleagues" (*Mind* 1977, p. 259) by declaring in a special footnote written just for Harre in the English version of *Erkenntnis für freie Menschen*, Footnote Number Four: "I have spent quite some time looking for the comments 'about women...' that so upset Harré. I could not find them. Am I blind or is he hallucinating?" Feyerabend, 1978, p. 131. Feyerabend, it seems, may well have been blind enough to such issues; yet for his part, Harré was surely hallucinating a deliberate insult though not the offensive implications that tend to follow in the wake of genial enthusiasm. For on p. 185 of the same book, Feyerabend offers a precise "aside" to use Harré's word – if perhaps not a "comment" as Feyerabend might have countered – to the effect that *Against Method* was nothing more than a kind of extended letter to Imre Lakatos, musing that he had instead "considered dedicating the book to three alluring ladies who had almost prevented its completion." Such an aside isn't meant to mean much – that's why it's an aside – but it is difficult to overlook. Indeed, contemporary feminist sensibilities more than endorse Harré's scruples.

16. Feyerabend (1989), p. 189, cites Mach's claim that research "cannot be taught" from Mach's *Erkenntnis und Irrtum*, Leipzig, 1917, p. 200.

17. That is, apart from the cascades of rhetoric poured off to confuse his readers. When such tricks succeeded, Feyerabend, in middling Viennese caprice, dubbed his readers illiterate. See Feyerabend, 1981, 1987.

18. See Richardson; for a recent review see Babich (1995).

19. In a characteristically circular assertion, H. Maturana, a contemporary scientist, notes that such an assumption constrains the use of the word science: "*das Wort Wissenschaft wird jetzt in der Regel nur noch auf Erkenntnisse angewandt, die mittels einer bestimmten Methode, nämlich der wissenschaftlichen, bestätigt wurden.*" in Watzlawick, P., Krieg, P. (eds.), (1991), *Das Auge des Betrachters: Beiträge zum Konstruktivismus*, Piper, Munich, p. 167.

20. For an account of the functioning of the former, see Ian Hacking's two books on probability and statistics (particularly 1990, *The Taming of Chance*, Cambridge University Press, Cambridge). For the latter see almost any robust or strong study of the sociology of knowledge.

21. For Toulmin, as a doctrine of systematic rationalism, that is, doctrinally "the trajectory of Modernity has closed back on itself, into an Omega; but experientially it has headed broadly upward." (Toulmin, 1990, 168) This parabolic reflex, a figure Toulmin employs to describe the harmonious tension of postmodern ambiguity, points to the increasingly essential growth of a "discriminating care for human interests" as distinguished from what Toulmin sees as the "scaffolding" of Modernity: namely its exclusively theoretical agenda. This is the hiddden agenda of modernity (at least one) and it is this theoretical project which entails the "separation of humanity from nature and [the] distrust of emotion" so characteristic of the phallogocentrism of received rational discourse of the sciences, of philosophy, and of theory in general.

22. Likewise, the erstwhile emblems of high-modernism/capitalism, nations and corporations, contine in newly diversified incarnations.

23. It is an apparently innocent, even touching conviction of many critical theorists that showing the modernist roots and modernizing vision and heart of postmodernism should somehow prove fatal to its exponents. This rhetorical faith tells us more about the conceptual limitations of routinely modern or enlightenment reasoning than it does about the genealogical limitations or core vulnerabilities of postmodern thinking. Beyond the clichéd charge of contradiction, the limits of postmodern

thinking may be better traced by questioning the authenticity or good faith of its claims to playfulness, and further, as I propose at the conclusion of this essay, by examining the challenge of "regaining" what Nietzsche, as a test of maturity, called the seriousness of a child at play. See reference and full citation in note 26 below.

24. Here we note the dovetailing of value judgments with an explicit reference to the market and hence a tacit qualification emblemmatic of a moribund capitalist economy, with no end, and no present alternative. Now an enthusiastically unqualified encomium, the saying 'the best money can buy' once might have suggested the limitation of market offerings together with a reference to the transcendent – qua that possibility that stands higher than actuality. This traditional reference to a transcendent involved an intangible handworking quality or contribution of the artisan's heart and dedication to other spiritual or personal values, underlining the distinction between the market and the soul. This quality of the soul, apostrophized not only by Kant as 'without price,' has been lost.

25. And those of us who feel liberated by our word processing programs, by our spread sheets, have a number of problems not least of which is that we have forgotten the point of an old story which Plato told in the name of an Egyptian myth concerning the Ibis-headed god of technology (and of magical advantage), Theuth, and an Egyptian king, the god's human counter, Thamus.

26. *"Reife des Mannes: das heisst den Ernst wiedergefunden haben, den man als Kind hatte, beim Spiel."* Nietzsche, Vol. 5 *Jenseits von Gut und Böse, "Sprüche und Zwischenspiele,"* 94, p. 90

27. I thank Holger Schmid for discussing this Nietzschean point with me and for his remark that the child plays exactly without irony. It may be recalled here that Nietzsche's Zarathustra named the child "Innocence," (*"Unschuld"*) but also *"Vergessen, ein Neubeginnen, ein Spiel, ein aus sich rollendes Rad, eine erste Bewegung, ein heiliges Ja-sagen."* Also Sprach Zarathustra, *"Von den drei Verwandlungen."* Nietzsche, Vol. 4, p. 31. Cf. *"Von Kind und Ehe,"* p. 90. In the postmodern context it is important to reflect that both self-consciousness and self-assertiveness destroy innocence.

II
THEORY OF SCIENCE:
QUANTUM DECONSTRUCTION

3 An Anti-Epistemological or Ontological Interpretation of the Quantum Theory and Theories Like It

Patrick A. Heelan

I[1]

I want to argue the thesis that the quantum theory needs to be given an ontological rather than an epistemological interpretation. By *epistemological*, I mean, relative to the cognitive content; by *ontological*, I mean, relative to the activity of representing the cognitive content. The quantum theory, I hold, describes phenomena as revealed through a process of empirical inquiry that is socio-historical in character rather than objective in the traditional sense. Such an inquiry is one fundamentally related to local communities and aims to describe something in the world from the point of view of a localized expert witness. Such a model of research implies a role in the scientific account for two non-classical freedoms, i.e., for social factors and for history. The ontological viewpoint I am proposing is inspired by traditions as old as Aristotle and as new as Heidegger.

By the terms "theories like it [the quantum theory]" I mean theories capable of articulating phenomena disclosed only by a kind of research having an essentially socio-historical dimension.

Among the technical terms used in this paper is the term "phenomenon"; by that I mean an individual entity in the world that is understood to have essentially socio-historical characteristics. I shall speak more below about its special relevance to the quantum theory. Most philosophical clarifications, however, will be postponed to section III.

II

Before plunging into the philosophical details of the analysis, I want to present a formal model for socio-historical theories of the quantum theory type. This section can be skipped for those less interested in the mathematical modelling of scientific theories.

I take the space of normalized ket-vectors (normalized rays in Hilbert space) to represent (make a formal model of) the research acts or states of any local suitably prepared data producing research community, which as above I call "the observer" or "the local observer." These are the local expert witnesses, either active in the performance of the inquiry or sophisticated onlookers and judges of the process. The subject matter of the inquiry is a local phenomenon named P. Its observable qualities, named Q, R, S, etc. are represented for the local observer by a set of data operators, Q, R, S, etc.; let the range of the data values of Q, R, S, etc. be respectively q_i, r_j, s_k, etc. where q_i, r_j, s_k, range over the eigen values of the data operators Q, R, S, in the Hilbert space of the research acts or states.

I suppose the following:

1. Research acts or states represented by the ket-vectors are given an ontological, not an epistemological interpretation.

2. The data operators, Q, R, S, etc. are representations of some appropriate space-time transformation group. The purpose of this is to give the phenomenon appropriate space-time properties, e.g., that its occurrence is essentially independent of where and when it occurs in the space-time of the scientific account.

3. Some pairs of data operators do not commute in the Hilbert space, e.g., QR and RQ may lead to different results according to the theory.

4. Measurement (or data judgement) results in a discontinuous change in representation from a superposition to an eigen state.

Let $|0>$ be the initial state of the local observer; this is specified by some past data judgements (data reports).

Let us further suppose:

5. There are differential equations of motion (analogous to the Schrödinger equation) that govern the change of state of the inquiry in the absence of new data judgements. If Kdt is the differential operator that represents the differential change in any state after the lapse of time dt; then, the research state or state of the inquiry becomes $(|0> + K|0>dt)$ at time dt in the absence of new data judgements.

Consider the quality Q and the data operator Q that represents it. Let the eigen states of Q – all mutually orthogonal – be $|q_1>$, $|q_2>$, etc. The vector $|0>$ can be expanded as a superposition of these eigen states:

56

$$|0\rangle = a_1(0)|q_1\rangle + a_2(0)|q_2\rangle + \ldots \tag{1}$$

where $a_1(0)$, $a_2(0)$, etc. are the complex normalized coefficients

$$a_1(0) = \langle q_1|0\rangle, \; a_2(0) = \langle q_2|0\rangle, \text{ etc.}$$

Let us further suppose:

6. The probabilities of occurrence of the values q_1, q_2, etc. under measurements of the quality Q represented by Q are equal to the absolute squares of the complex normalized coefficients, i.e., $|a_1(0)|^2$, $|a_2(0)|^2$, etc.

Let there be a new measurement of Q and let the new realized value of Q be q_1. The research state following this individual Q-measurement can then be represented by the eigen state

$$|q_1\rangle. \tag{2}$$

Such a new measurement and datum judgment are like a measurement in quantum mechanics, they change the state of the inquiry discontinuously from one represented by a superposition (1) to one represented by the single eigen state of Q, (2), associated with the value q_1. The probability of occurrence of q_1 is $|a_1(0)|^2$. The change from $|0\rangle$ to $|q_1\rangle$ represents the constitution of the datum represented by $Q = q_1$.

The statistics generated in this way from coefficients in the expansion of a prior (pre-measurement) state in terms of the eigen basis of a particular data operator or, more generally, in terms of the common eigen basis of a maximal set of commuting data operators are non-classical in character.

III

The formal structure just sketched is the core of the quantum theory, and any theory with such a core, I call a theory of the quantum theory type.

The Hilbert space of ket-vectors is usually interpreted objectively from an epistemological viewpoint, i.e., the rays or ket-vectors are taken to picture objective states of a real phenomenon independently of human knowledge and measurement. Measurement, however, and the non-commutivity of some data operators are problems for the quantum theory. Measurement is associated with a discontinuous change in the system representation called "the reduction of the wave packet" and what this represents is something of an unsolved mystery for those who assume that the quantum theory has an objective interpretation. The same can be said of non-commutivity of data operators; what is measured first in this case destroys correlations that affect the second measurement, and vice versa.

I now want to say that the mysteries of quantum mechanics can be resolved by giving theories of the quantum theory type an ontological interpretation, i.e., by taking them to refer to the essentially socio-historical processes involving local expert witnesses in which the phenomenon becomes disclosed. This disclosure occurs differently when taken over different socio-historical paths; the phenomenon takes "flesh" in the world differently because its "flesh" is determined only as a consequence of decisions taken by local and historical communities of expert witnesses. (The metaphor of "flesh" is borrowed from M. Merleau-Ponty.) To put it another way and using a different metaphor: the phenomenon always makes it appearance in some "dress", but in which of its possible "dresses" it makes an appearance is a consequence of human decisions made locally. (There is, of course, no such thing as an "undressed" phenomenon; the metaphors of "flesh" and "dress" are introduced to substitute for Aristotle's "quality" while relativizing it to human communities and history; cf. Heelan, 1989.)

Firstly, we can give an account of the essential unity of a phenomenon as understood and experienced by local communities since all presentable data are united formally by a single theory of the ontology of acts of understanding and experiencing the phenomenon.

Secondly, superposition states represent research states of the local community at the moment when a decision is being called for and before it is executed in the real world. (Choosing to represent a state as a superposition state relative to a data operator Q is, I take it, the first step toward a decision to measure the quality Q). The "reduction of the wave packet" represents a new ontological state of the research community and this is a real change in the world.

Thirdly, a phenomenon is characterized by path dependent data. For example, for the quantum phenomenon, a change in the order of some pairwise decisions or measurements, such as momentum and position, or up-down and left-right spin, would (according to the theory) yield different sets of data irreversibly. The elements of each pair, though mutually exclusive in their historical realizations, belong nevertheless to the definition of the same phenomenon insofar as this is full of real historical potentiality. We could speak of the same phenomenon as "fleshed out" by and for a local research community by the sequence of decisions made by that community; it is this sequence that determines the historical path of its evolution in the local world of that local community. The phenomenon remains, however, throughout its evolution the *identical* phenomenon.

While the epistemological account generally supposes that presentations of data are not path-dependent, i.e., uninfluenced by different possible local historical paths that the inquiry could take, the ontological account of data constitution does confront the ontology of social-historical time as constitutive of human understanding and experience, and the sequential order of local human decision-making as branch-points of real-world novelty. In the model here proposed, data production is path-dependent. Different local and historical communities can prepare and recognize the same phenomenon, but they will find it "dressed" or "fleshed out" differently – exhibiting the phenomenon's social and historical dimensions.

The ontological interpretation implies a role in the scientific account for two non-classical freedoms, i.e., for social factors and for history. Social factors enter at decision points where local communities decide what superpositions are to be broken by measurement. History enters through the path-dependence of the data that trace the life history of the real phenomenon. Although both of these freedoms are real and can lead to real novelty, they are constrained by the essential unity of the phenomenon. This essential unity is defined by the quantum theory type model. As long as the data operators and their commutation relations remain fixed, the social and historical variability is constrained by the unusual "quantum" statistics yielded by the model. These statistics are derived from the expansion of a prior (pre-measurement) state in terms of the eigen basis of a particular data operator (or more generally in terms of the common eigen basis of a set of commuting data operators); it is from these expansions and superpositions that frequency functions are derived.

Theories of the quantum type set limits to the social and historical variability of data; these limits are set by the non-classical statistics of the model. Data sets relevant to one and the same phenomenon displaying essential historicity and social variability should then be tested by these new "quantum" statistics. This conclusion should be empirically testable in areas such as the social sciences which are deeply affected by historicity and society and where, one thinks, theories of the quantum type may apply.

I do not wish to provide at this point a lengthy discussion as to how secular changes in the ways the phenomenon is "dressed" over longer historical periods can be reflected in models of the quantum type. Some secular changes are evolutionary, some revolutionary. The quantum model has the potentiality of showing how changes in science often described as revolutionary can be seen as evolutionary, i.e., by changes making the old theory more comprehensive while preserving continuity with its past. In theories of the quantum type, new data operators can be added to the Hilbert space, leaving the rest of the structure intact. This is permitted by the infinite dimensionality of the ket-vector representation, a quality which gives it infinite potentiality for further specification. New data operators introduce, of course, new commutation relations with the old data operators and among themselves. The introduction of spin operators to quantum mechanics in this way was evolutionary, because it enlarged the representation of what was measurable from quantum phenomena with zero spin to phenomena with higher spin. The old model has become the ground floor of the new edifice.

Revolutions in science, which result in different data being reported in the same categories or for which new and different categories replace the old ones for the analysis of reportable data, raise the specter that scientific inquiry, however careful, may not provide understanding of the ontology of scientific phenomena, i.e., of these phenomena as real in the perspective of Being. The reason for this conclusion is that most of the discussion about scientific revolutions has supposed that a scientific phenomenon is just a word for data as elements of a theoretical synthesis, regardless of whether or not the data synthesis implies an ontological interpretation of Being, i.e., the existence of something like a substance that is "dressed" both by the data and the synthesis and is the *subjectum* revealed in both.

It is now time to return to the philosophical background that is implied by everything I have said and to attempt a proof by clarification of what has been said.

IV

Many of the sciences, but especially the social sciences, rely on methods of statistical analysis originally developed for the physical sciences; in these methods it is assumed that data are in principle *objective*. By this I mean particularly two components, 1. that what a datum is (how it should be described) does not depend on the local interests people may have in measuring and recording it, i.e., the descriptive categories for data and phenomena are not derived from the local character of the scientific community but are rather in Newton's terms "universal" qualities (Newton, 1962, vol.2, "Rules of Reasoning in Philosophy") and 2. that whether or not a datum is recorded (i.e., a data judgment is made) does not change the course of things in the world, i.e., the observation and recording of data and the phenomena they reveal leave the world unchanged. In this classical view scientific inquiry is a view from outside the world, disengaged from the course of events in the world.

Of the two components of objectivity stated above, the former is concerned with the categories of objective knowledge, let me call it an *epistemological principle*; the latter is concerned with the being of the world, let me call it an *ontological principle*. It is well known that quantum physics has undermined the latter of these principles, since one cannot make a measurement without changing the state of the measured phenomenon; on the terms of the interpretation given above, quantum physics also undermines the former; with respect to both these principles, it is not clear what should replace them. These objectivity principles have been challenged in recent years by philosophers, sociologists, anthropologists, and historians of science, all of whom speak on the dispute from a common perspective, and this perspective is inadequate. It is inadequate because it is epistemological (in the classical sense), i.e., concerned with the relation between the domains of the mental (or more generally, the representational) presumed known and the physical presumed derivable from the representation. It is clear that this is a dead end, for epistemological problems are not resolvable without a prior ontological analysis.

What is the meaning of ontology? and what is the ontology of an act of research, i.e., of acts in search of understanding what we experience? Understanding and ontology are linked. Inquiring, searching, understanding are, as Aristotle said in the *Metaphysics*, of the defining essence of being human. The domain of this activity is traditionally called Being. Acts of understanding relate people to the world by recalling where they are, where they have been, and where they might want to go. Each act of human understanding foresees possibilities and envisages choices against the background of the World; World in this sense is Being. Heidegger puts it well when he says that to be human is to-be-There-in-the-World, to be *Dasein* (Heidegger, 1962, p. 27, and in general, sect. 32; see also Heelan, 1983 and 1989). By Being and World, I mean, not a collection of ready-made things and events, i.e., of beings and local worlds, but the background essential to human life that makes all things and events possible and understandable. Being is all there is and *can be* and human life is ultimately concerned with nothing but this.

Heidegger made a distinction, the rudiments of which are also found in Aristotle, between "ontic" beings and "ontological" Being; this he called "the ontological difference." (Heidegger, 1966) Ontic beings are given as ready-made things and events distributed in objective space and time and belonging from the theoretical viewpoint to the inventory of a local Euclidean world – this is reality as it is usually taken in modern philosophy – while ontological Being is the common essential background of human life which makes ontic beings understandable and which, when interpreted by human life, confers reality on local worlds. Aristotle and Heidegger are representative of a tradition that addressed the activity of human understanding and research first and foremost as ontological, and only secondarily as the ontic. Pragmatism tries to escape the rigidity of modern philosophy's ontic framework by introducing human life and culture into the specification of the world (Dewey), but such specifications fail to raise the ontological dimension, and Pragmatism consequently fails to be able to justify itself on other than instrumentalistic grounds.

The activity of understanding then is to be defined as the ontological activity of constituting local beings as known within the ontological horizon of Being. Note that since understanding (the activity of) understanding does not presuppose prior epistemological principles (taking these in the modern sense to state how one can move in a justifiable way from mental representations to knowledge of reality), it avoids the basic *petitio principii* of an epistemological starting point. One of the functions of such *constitution* is to construct and use representations; what these are and how they are used is then to be studied in this connection.

For Aristotle and, more generally, prior to Descartes and Locke, epistemological questions were formal, regional, and local in character, subsidiary to the ontology of knowledge. After Descartes and Locke, however, the general cultural consensus characteristic of modern times was formed, that science provides the single, true, and privileged account of Nature and that such an account ought to replace all other accounts of Nature. Modern philosophy turned to epistemology largely because of the dissonance between Nature as (ontically) pictured by science and Nature as assumed for the purposes of human life. The new and emerging scientific picture of Nature was bereft of those sensible qualities and feelings, moral purposes, and social organization that constituted the arena – the World – of human life; it had of itself no "meaning," i.e., no human, social "meaning" (whatever other meaning it had was in the cold eye of an impersonal God). In this cultural transformation it was inevitable that a new branch of philosophy would grow up, a general epistemology, which was both the science of scientific knowledge and claimed to be the foundation of all philosophy.

When epistemology moved to center stage, problems were created that could at first be postponed. How can knowledge build bridges between representations, such as scientific theories, diagrams, or data reports, and reality without presupposing a prior grasp of reality? The changes that have taken place in science since the 17th century have brought new and very different models of scientific knowing and with them confusion to those who believed that science provided unchangeable and privileged paradigms of objective knowledge. Certainly, it was the advent of quantum mechanics that dealt the severest blow to the view that scientific knowledge is objective. It is to solve these problems that I turn to an ontology of the activity of scientific research.

Between an epistemology and an ontology of scientific research, there are significant differences in starting point and method. In the *first* place, we drop the Cartesian supposition that whatever functions as a representation can itself be known and judged by internal criteria and we return instead to the Aristotelian position that representations (or *species*, as they were called) are not generally known in themselves but only in what they make known. We work from the principle that whatever scientific representations do, they do it only as a function of what human understanding is.

In the *second* place, we suppose that real or evident knowledge activity has somehow antecedently been identified and described, but not by an epistemological inquiry in the traditional sense, i.e., concerned with the justification of mental or other representational contents. Such a starting point can be identified by using the resources of other philosophical traditions sometimes included under epistemology. In particular, phenomenological inquiry has immediate reference because it is concerned with the ways Being is given critically and evidentially to human understanding. For E. Husserl, M. Heidegger, M. Merleau-Ponty, and the phenomenological tradition generally, phenomenology is not about representations and their validity, but about "*die Sache selbst*," i.e., about what is represented with evidence in the act of knowing (cf. Husserl; Heidegger, 1962; Merleau-Ponty). And yet, any reader of the phenomenological literature will notice that until recently there has been a kind of hostility towards scientific research except towards mathematics, and no studies of such an important activity as data preparation, recognition, and reporting in laboratory environments. This feature of the internal history of phenomenology is all the more surprising since Husserl, the founder of philosophical phenomenology, himself was a core member of the charmed Göttingen circle of mathematician-physicists who, during "The Golden Years" of Göttingen at the beginning of this century, set the scientific agenda for physics for the twentieth century. (Heelan, 1987)

In the *third* place, we take the paradigmatic fulfillment of scientific knowing to be the recognition of occurrences in the world that science speaks of, i.e., of scientific data. Data constitution is the starting point and the central philosophical question of an ontological inquiry into science.

Data production and analysis is a part of the experimental, i.e., laboratory, side of science. Compared to the interest philosophers have shown in theories, interest in experimental work has been small until recently.[2] Recent work, however, has failed to clarify the notion of data and to address the central philosophical problems of the phenomenology and ontology of data constitution (see Heelan, 1989).

I shall first present a clarification of the notion of *data*, and then return to the proposal discussed in Section II. We begin with a phenomenology of data observation: every datum is in relation to a phenomenon and in relation to a local suitably prepared community of data producing observers (Heelan, 1989). By *community*, I mean, a group that monitors the activity of its members. By a *local* community, I mean, one that exists in a certain place and time, not an ideal, universal, or global community. By *suitably prepared*, I mean, sharing an expertise and equipped with the appropriate instruments. By *data producing*, I mean, preparing, recognizing, and reporting data in question. (This community is the same I earlier called a "local community of expert witnesses.") In sum, I am speaking of the activity of research, particularly of laboratory science. I shall use the terms "observer" or "local observer" for such a local suitably prepared community of data producing observers.

For a philosophical perspective, the local observer must be raised above the everyday attitude (i.e., ontic, instrumental, technical, or other) to become critically self-aware as *Dasein,* i.e., as understanding Being; ideally, philosophers of science occupy the position of such a local observer. An antecedent phenomenological critique that is sensitive to the experience of doing laboratory science has for its goal the initiation of such a philosophically located community. It should be noted, however, that the initiation of such a philosophically located community is merely preparatory to the further work of interpretation which is philosophy. Interpretation against the background provided by the textual and other resources of the philosophical community constitutes philosophy.

Considered phenomenologically, every datum is *for* an observer and *about* a phenomenon. The datum is some (real) appearance of a phenomenon to an observer; the datum judgment then is always about some phenomenon revealed through a datum, e.g., "The energy of this (just arrived) electron is 5 Mev." A (real) appearance of a phenomenon is often called a "profile" of the phenomenon, *having energy of 5 Mev* is a profile of the (just arrived) electron.

The phenomenon is the existential unit and essential invariant to which a multiplicity of data can be ascribed by the observer. The "about-ness" or "of-ness" of a datum implies that every datum is *intentional*, i.e., that every datum is an appearance *of* something else that is individually present.[3] The intentionality of a datum is threefold. 1. In the first place, a datum is not known for its own sake, but for the sake of something else that it reveals as present in the ambience of the observer; we call this *the phenomenon*. 2. In the second place, the datum judgment is synthetic (joins predicate [P] to a subject [S], e.g., S is P), since it is the nature of a phenomenon (S) that, while remaining essentially the same, it is capable of showing itself to an observer under many different appearances (P's); the phenomenon then is the subject (*subjectum*, see below) to which is attributed a possible multiplicity of connected data. 3. In the third place, since the phenomenon in fact endures between appearings to observers, the phenomenon is more than any data accumulated about it and more than the law of data synthesis. Aristotle, though far removed from the sophistication of modern science, saw this point clearly and called what is so posited "substance" or "*subjectum*." This position re-asserts something that modern philosophy denied, namely, that human understanding has an intuition of existence or being that goes beyond the sensibles or the categories. It is this insight we are trying to recover when we say that the datum judgment affirms in addition to the synthesis (e.g., S is P) the phenomenon as a being beyond both the data and the synthesis; more precisely, the synthesis is not just one of correlations among data, but of attribution to a common *subjectum*.

In relation to the research activity of any local observer, the *epistemological datum* is the content of what is objectively represented when a datum judgment is formulated as a report through the medium of some language and some theory. Data reports are offered in the form of numerical indices or values estimated by instrumental measurement and attributed to the qualities possessed by the phenomenon according to the theory in use (cf. Ackermann). Data reports are both *concrete,* denoting what has appeared to observers in space and time, and *abstract,* connoting the essential qualities of a kind that can occur again and again in observation or be prepared in standard ways under standard circumstances for observation. The language of data reports is, on the one hand, sensibly realistic and environmentally worldly, and on the other, loaded with numbers and theoretical vocabulary. This creates a paradox but it is not here my intention to resolve this; the topic has already been addressed in a recent paper (see Heelan, 1989).

Data reports are the product of constitution and, once made, all reference to the process of constitution and representation is dropped; in contrast, the *ontological datum* (taken in the context of the *observer-being-in-or-coming-into-the-presence-of-a-datum*) retains this reference.

In the formation of an epistemological datum, a purge of locally irrelevant but ontologically significant factors may have, indeed usually has already, occurred, as in quantum mechanics. *It is this feature that suggests a comparison with the quantum mechanical model of a transition from superposition states (before measurement) to a pure state (following the datum judgement).* Measurement in quantum mechanics is accompanied by a loss of information (an increase in entropy). The new or "quantum" statistics of constituted data exhibit strange features familiar to quantum theorists among which the most infamous are the limits set by Bell's Theorem (according to which quantum states fail to be "local" and "causally connected", i.e., they can retain their space-time identity over large separations and produce correlations not explainable by point to point transfer of information as classically envisaged). The constitution of the datum in the kind of theory we are considering is then the ontic realization of one possibility, executed at the expense of some excess of ontological possibilities that are lost irreversibly in the process; this irreversible change in the real course of historical events brings about the above mentioned increase in entropy.

What this original excess of ontological possibilities is will need a thorough discussion. It may be useful to envisage it provisionally in the following way: on the one hand, it is a real defect in the current research state relative to what is necessary for a determinate object to appear to a local observer or for a determinate data judgment to be made; certain switches (as it were) have to be turned on. On the other hand, this real defect can be envisaged positively as an *excess of ontological possibilities* of such a kind that, for the inquiry to be completed, certain switches are, to the contrary, (as it were) turned off. The point I want to make in this provisional account is that, at any research stage of an inquiry, decisions may be made and executed which suppress superpositional states, giving the inquiry an historical path.

The notion of a path, i.e., an *historical path* tracing a sequence of human decisions, has been introduced earlier. By an *historical event*, I mean one within the life World of a particular human community loaded with a sense of the community's lived past and of decisions to be made for the future. It is not the case that every historical event is also an event of a scientific kind (a scientific datum), but when the local community is one of expert witnesses, then the scientific data produced by that community are also historical events in relation to that community.

V

Finally, some conclusions: theories of a quantum type are suggested wherever the phenomena under study have an essentially socio-historical dimension on account of which data appear only as related to a specific path, i.e., 1. to specific local communities of expert witnesses, such as different laboratories, different historians, different nationalities, etc. and 2. to the sequencing of the phases of the inquiry (the order in which decisions are made and executed). Examples of the former are presently used to cast doubt on the existence of a single phenomenon; while examples of the latter are now turning up in the sequencing of questions in questionnaires, of pedagogical materials in teaching, of stages of economic development, etc.

References

Ackermann, R. J. (1985), *Data, Instruments, and Theory*, Princeton University Press, Princeton.
Bernard, C. (1957), *Introduction to the Study of Experimental Medicine*. Greene, H.C. (Trans.). Foreword by Cohen, I.B. Dover, New York.
Cartwright, N. (1983), *How the Laws of Physics Lie*. Oxford University Press, Oxford.
Collins, H. (1985), *Changing Order: Replication and Induction in Scientific Practice*. Sage, Beverly Hills.
Crease, R. and Mann, C. (1986), *The Second Creation: Makers of the Revolution in 20th Century Physics*, Macmillan, New York.
Dewey, J. (1960), *The Quest for Certainty* Putnam, New York.
Dirac, P. A. M. (1967), *Principles of Quantum Mechanics*. 4th Ed. Rev.

Clarendon Press, Oxford.

Fleck, L. (1979), *Genesis and Development of a Scientific Fact*, Trenn, Thaddeus (ed. and trans.), Bradley, Fred (trans.), Merton, Robert (ed.). Foreword by Kuhn, T.S. University of Chicago Press, Chicago.

Franklin, A. (1986), *The Neglect of Experiment*, Cambridge University Press, Cambridge.

Galison, P. (1987), *How Experiments End*, University of Chicago Press, Chicago.

Hacking, I. (1983), *Representing and Intervening*, Cambridge University Press, Cambridge.

Heelan, P. A.(1983), *Space-Perception and the Philosophy of Science,* University of California Press, Berkeley and Los Angeles.

Heelan, P. A. (1987), "Husserl's Later Philosophy of Science," *Philosophy of Science*, 54: 368-390.

Heelan, P. A. (1989), "After Experiment: Research and Reality," *American Philosophical Qrtly* 26: 297-308.

Heidegger, M. (1962), *Being and Time.* Macquarrie, J. and Robinson, E. (trans.), Harper and Row, New York.

Heidegger, M. (1966), *The Essence of Reasons*, Northwestern University Press, Evanston.

Holton, G. (1973), *Thematic Origins of Scientific Thought: Kepler to Einstein*, Harvard University Press, Cambridge.

Holton, G. (1978), *Scientific Imagination: Case Studies*, Cambridge University Press, Cambridge.

Hughes, R. I. G. (1989), *The Structure and Interpretation of Quantum Mechanics*, Harvard University Press, Cambridge.

Husserl, E. (1970), *The Crisis of European Sciences and Transcendental Phenomenology: An Introduction to Phenomenological Philosophy*, Carr, David (trans.) Northwestern University Press, Evanston.

Ihde, D. (1979), *Technics and Praxis.* Reidel, Dordrecht.

Kuhn, T. S. (1962), *The Structure of Scientific Revolutions.* 2d ed., University of Chicago Press, Chicago.

Kuhn, T. S. (1977), *The Essential Tension: Selected Studies in Scientific Tradition and Change*, University of Chicago Press, Chicago.

Latour, B. (1987), *Science in Action*, Harvard University Press, Cambridge.

Latour, B. and Woolgar, S. (1979) *Laboratory Life: The Social Construction of Scientific Facts*, Sage, Beverly Hills.

Merleau-Ponty, M. (1962), *Phenomenology of Perception*, Smith, C. (trans.), Routledge and Kegan Paul, London.

Newton, I. (1962), *Mathematical Principles of Natural Philosophy and His System of the World.* Motte, A. (trans.), Cajori, F. (ed.), California University

Press, Berkeley and Los Angeles.

Pickering, A. (1984), *Constructing Quarks: A Sociological History of Particle Physics*. Chicago: University of Chicago Press.

Rouse, J. (1987), *Knowledge and Power: Toward a Political Philosophy of Science*, Cornell University Press, Ithaca.

Shapin, S. and Schaffer, S. (1985), *Leviathan and the Air-Pump: Hobbes, Boyle, and the Experimental Life*, Princeton University Press, Princeton.

Notes

1. Further aspects of the topic of this 1991 paper were subsequently developed and submitted for publication as a paper, "Quantum Mechanics and the Social Sciences: After Hermeneutics," in *Science and Education*, Matthews, M. (ed.), (1995) Kluwer, Dordrecht, and in *"Herméneutique de la Science Expérimentale: La Mécanique Quantique et les Sciences Sociales,"* prepared for a seminar at Cerisy-la-Salle, Paris, September 1994, and to be published in its proceedings.

2. The recent interest in experimentation grew out of the seminal studies of L. Fleck (1979), G. Holton (1973, 1978), T.S. Kuhn (1962, 1977), C. Bernard (1957), and others. Among the more significant books on this topic recently published, some are by philosophers, such as R. Ackermann (1985), N. Cartwright (1983), and A. Franklin (1986), I. Hacking (1983), P. Heelan (1983), D. Ihde (1979), J. Rouse (1987); some by historians, such as R. Crease and C. Mann (1986), P. Galison (1987), S. Shapin and S. Schaffer (1985); and some by social scientists, such as H. Collins (1985), B. Latour (1987), A. Pickering (1984), B. Latour and S. Woolgar (1979).

3. With respect to observers, data perform a role analogous to what *sensations* would perform in classical theories of perception, *if these were ontological theories rather than epistemological theories;* they mediate the real presence of an object to an observer, but with the difference that instruments play the part of sense organs.

4 The Play of Nature: Experimentation as Performance

Robert P. Crease

Introduction

It is now common for philosophers of science to proclaim the discipline to be in a state of crisis.[1] Canonical authors have lost their authority; nobody is sure which way to turn; a variety of modest initiatives are proceeding in a number of different directions. Throughout this period of crisis, however (which may have lasted from one to four decades, depending on the diagnosis), philosophy of science has had the air of a diseased physician struggling futilely against a raging, self-inflicted malady. The physician is so consumed by the struggle that hardly any time is left for a glance at the "patient," science itself, who meanwhile is in quite sound health, utterly oblivious to and unaffected by the paralysis of its alleged general practitioner. To recommence practice, two things must happen: first, the physician must self-administer the right therapy with the right set of tools, and second, the physician must begin to look again at the patient.

One important case of where philosophy of science is in dire need of tools and a real examination of the patient is experimentation.

Experimentation is a form of research that aims to bring a phenomenon into material presence in the service of an inquiry. Experimentation can mean either the uncovering of new and hitherto unknown phenomena or the continued study of known (already recognized) phenomena prepared in the laboratory. In successful experimentation, phenomena show themselves as "the same" – and not as glitches, errors, epiphenomena, or artifacts of the machine – to first-hand observation by readable technologies. Experimentation, in short, is not merely a *praxis* – an application of some skill or technique – but a *poiesis*; a bringing forth of phenomena. One can mean by experimentation either the execution of a particular kind of performance, or a particular kind of inquiry which includes the execution of experimental performances as its principal part. Experimentation characterizes science in that taking objects and events as instances of phenomena, and the desire to extend knowledge of them, is the *cultural attitude* of science.

From this brief description it is already apparent that concepts essential to a philosophical investigation into experimentation include *inquiry, invariance,* and *interpretation.* Accounts of these concepts which can be suitably adapted to the context of experimentation can be found in the works of Dewey, Husserl, and Heidegger, respectively.

Dewey, Husserl, Heidegger

Dewey's account of inquiry views it as arising from the desire to reconstruct problematic situations, which when successful alters inquirer, environment, and forms of interaction between the two. Successful inquiry results in a more assured, enriched, and deepened experience and engagement with the environment. Inquiry is "technological;" discoveries are not additive but transformative of the environment for human purposes to supplement nature. The productions arising from successful inquiry have primacy over both theory and practical knowledge, each of which derives its significance from such production; skill divorced from it is tedious and repetitive, while theory separated from it is arbitrary and whimsical. Yet successful production promotes the illusion that its fruits preceded it, giving rise to the "philosophic fallacy;" the taking of results of inquiry as existing prior to it, whereas in fact inquiry creates rather than recalls these results. Experimental inquiry introduces changes in the world to discern relations with other changes. In Dewey's valuable perspective, experimental activity is primarily concerned with performance of actions *in* the world rather than with confirmation of hypotheses *about* the world. This perspective allows us to address experiments as concrete physical events, as implicated in a social dimension, as belonging to an interminable rather than terminable inquiry, and as having philosophical implications in preparing a purgation of the representationalist view of knowledge in science, its final bastion.

Husserl's account of invariance provides us with an account of what appears in and through experimentation. When successful, experimentation involves the appearance of *phenomena*; experiments *prepare* phenomena for study. Phenomena appear through *profiles*, which one depending on the relative positioning of observer and observed. The regularity of profiles under passive and active transformations is the *invariance* of the phenomenon. An invariant names an identity of structures in subject and object; the noetic-noematic correlation.

Invariants entail *horizons* of possible profiles, given together with phenomena; in exploring phenomena in inquiry, horizons are constituted, filled in, revised, and extended. Horizons may be *internal* or *external*. One acquires confidence a phenomena exists and that it is accurately represented through an ability to move from profile to profile. Expectations may be *fulfilled* or not, for phenomena can reveal themselves in new and unanticipated ways, and their invariant structure may have to be adjusted accordingly. Crucial here is Heelan's *Ansatz* that scientific entities are phenomena accessible to perception via *readable technologies*, with perception linked with the model used to represent the object by instrumental praxes.[2] Phenomena of science then share the same essential structure as natural phenomena; they are apprehended in experience as invariants given through ordered (modelable) sets of profiles in a way constituted by the process of scientific research, theoretical and experimental. A first implication of this *Ansatz* is that criteria of objectivity for scientific entities are the same as that of life-world objects. The pertinent question is: Does the object exhibit an invariance through its appearances in a sufficiently varied multiplicity of ways and under a variety of conditions? A second implication involves the correlativity of noesis and noema in scientific phenomena. Like profiles of other phenomena, those of scientific phenomena change through different "positionings" of observer and observed, and with different external horizons different descriptions can count as "fulfilling" a profile. In a certain equipmental context, a particular measurement may fall into a certain realm of acceptability, and count as indicating the fulfillment of a certain profile of an electron. In a different context, a larger or smaller range of numbers will count as acceptable. But there is always the possibility that phenomena can present themselves in totally new and unexpected ways. A third implication involves a *dual semantics* for experimenter and theorist. For the experimenter, 'electron' is a real phenomenon, a piece of *material ontology* involved in causal explanations; the real presence of electrons in the instrumental set-up is causally involved in the events that take place there. For the theorist, 'electron' is an abstract term, part of a *formal ontology* involved in nomological explanations; the theorist delivers an abstract model for the phenomenon consisting of a set of equations.[3] This dual semantics is crucial for understanding science. If all data were theory-laden, experiments could not constrain theory because data from the start would be unintelligible without theory and could not therefore be used to test it; what data did appear would all be contextually legitimated. The independence of experimentation and theory exists in the form of this dual semantics, with theory developing a syntax for representing performances, and experimentation describing the performances themselves using this representation.

Heidegger's account of interpretation, or "the development of the understanding," sees it as not involve a reaching out from a subjective interior to an

71

objective exterior, but a development of a familiarity or understanding which we always already have. All understanding is finite, and takes place within specific cultural and historical traditions; interpretation is the development of a finite and already determined understanding. Inquiry involves development of the under-standing via moving in the *hermeneutical circle*. Even the planning, execution, and witnessing of actions in experimentation involves a making-explicit of what one already understand, a making-explicit that involves assuring, enriching, and deepening one's involvements and expectations. The hermeneutical circle is at work in each of the three dimensions of performance I shall mention below: in presentation in the coming-to-appearance of new phenomena (one brings new things into being only by coming to anticipate their appearance); in the representation of new phenomena (which is done only thanks to an anticipation of the profiles represented); in the recognition of new phenomena (one only recognizes the novel by being already familiar with it).

Surely these three accounts – of inquiry by Dewey, invariance by Husserl, and interpretation by Heidegger – are so disparate that they cannot be thrust together without doing injustice to the phenomenon, experimentation, that I would use them to understand. But the fact that such a mixture of differing sorts of philosophical discussions *seems* eclectic is an indication of how deficient is our current philosophy of science. No one of the three figures mentioned has all the elements needed for a satisfactory account of experimentation. The Deweyian perspective needs an appreciation of the experience of the actual presence of scientific entities in experimentation, and a greater sense of invariance. The Husserlian perspective needs to be supplemented with an account of how perceiving an object, scientific or otherwise, within its external horizon is a process of fulfillment or interpretation. The Heideggerian perspective needs to acquire a positive account of science that sees its activity as disclosing more than an abstract, theoretical description but new phenomena, and that does not see science as derivative with respect to other kinds of activities – that sees the aim of science as world-building rather than world-denying.

Argumentative analogies

To show the place of these three disparate pieces in an account of experimenta-tion, I shall use an *argumentative analogy*, to be distinguished from metaphor. In a metaphor, a secondary subject is used to point out certain features in a principal subject but without asserting an explicit shared deep structure ("Man is a wolf," to use Max Black's famous example); priority belongs to the principal subject. In an argumentative use of analogy, involving a point-by-point comparison between one thing and another, the priority of the two terms is reversed. The secondary term, suitably adapted, is used to bring to bear a single, organized, already articulated perspective on the other term; it becomes the technically correct, albeit sometimes slightly neologic, term. The argumentative analogy is thus a tool whereby a structured set of relations present in one area is introduced into another.

Argumentative analogies are a major tool of scientific thinking, in play whenever a set of equations is taken from one context and adapted to another. Consider the development of wave mechanics. Waves originally referred to a state of disturbance propagated from one set of particles to another. By what amounts to an argumentative use of analogy, electromagnetic radiation came to be treated as waves, subject with amendments to the body of knowledge called wave mechanics; electromagnetic radiation was the principal, waves the secondary subject. By another argumentative analogy, waves were introduced into quantum mechanics, with the primary term complex functions whose solutions were not themselves measurable but used to provide probability distributions. Each time, wave equations were extended to new domains each time and the concept of wave changed; we recognize more kinds of waves now.

The point is sufficiently important to warrant another example. Consider the genesis of the theory of beta decay, a form of radiation involving the emergence of an electron from the nucleus, by Italian physicist Enrico Fermi. Accounting for the phenomenon had baffled Fermi's contemporaries to the point where, in 1932, Danish physicist Niels Bohr seriously considered the possibility that the law of conservation of energy was violated in this process. At the end of 1933, Fermi developed a theory of beta radiation largely, as he puts it in the abstract to his first paper on the subject, "in modo analogo" to the theory of electromagnetism.[4] The creation and annihilation of electrons in beta radiation was treated as analogous to the creation and annihilation of photons in quantum jumps between electron orbits in the theory of electromagnetic fields. An analogous Hamiltonian function was chosen, and the interaction of particles and fields in beta radiation was treated as analogous to that between charged particles and fields in electromagnetic theory. Fermi introduced a new fundamental constant of nature – G, the Fermi constant – to play a role in the new theory analogous to the role played by the charge of the electron in the theory of electromagnetism. The result provided physicists with a structure for understanding beta radiation which tied together its various hitherto puzzling aspects and which pointed to an existing system of procedures, techniques, and equations (those associated with electromagnetism) to use in further explorations of the phenomenon.

Fermi's account proved incomplete. For instance, he had chosen to treat in detail the vector form of the interaction Hamiltonian because of the analogy with electromagnetism; George Gamow and Edward Teller, however, showed that scalar and tensor interactions were also possible, and their work was incorporated into the theory.

Was Fermi's method really necessary? It was, after all, "only" an analogy. Perhaps he could have started from scratch to develop his account of beta radiation, and begun with something other than electromagnetism. But what would have been the point? A body of understanding was available that, suitably adapted, revealed its utility for developing an understanding of the phenomenon.[5] Despite its genesis in analogy, Fermi's theory remained virtually unchanged through a quarter century of one of the most revolutionary periods of atomic physics.[6] Someone may object that the analogically-developed theory succeeded only because a real similarity exists in nature between electromagnetism and beta radiation. *But that was precisely what had to be proved.* The development of the analogy was what allowed that "real" similarity to disclose itself the first place, and if it needed to be adjusted or denied in part, as Gamow and Teller showed, that was not a refutation of the analogy but further disclosure of the phenomenon. This is the kind of process that motivated Jeremy Bernstein to write that, "It is probably no exaggeration to say that all of theoretical physics proceeds by analogy."[7]

Argumentative analogies are also an important tool of philosophy, when approaches and methods are transferred from one field of study to another. Analogy is indeed practically a universal tool of thought, a manifestation of the Vichian principle that human beings understand the unfamiliar by the familiar.[8] Analogies are useful as what Wilshire calls "physiognomic" devices, allowing us to reorganize our perception of phenomena. Analogies are of help in sketching out the aspects we should inquire into, and the kinds of outcomes that would satisfy us.

The attempt to consolidate and coordinate what is known (philosophically) about experimentation can proceed via an argumentative analogy in which performance is the secondary and the acts by which scientific entities become perceptible and present (experiments) the principal subject. In arguing for the analogy between experimentation and theatrical performance, I am saying that the two kinds of activities are each forms of performance and share a similar structure on an abstract level, and mean to exploit this similar structure to extend our knowledge of the structures of experimentation.

One searches in vain in traditional textbooks of the philosophy of science, for instance, for discussions of the significance of choosing the right laboratory, of the political process needed to shepherd a proposal through a program committee, of the presentation of one's results, of getting carried away by the theatricality of it all and deceiving oneself about results, of selecting a team, of intrateam tensions, of the discovery process itself (often dismissed by traditional philosophy of science as only of psychological interest), or of the joy that accompanies creation or discovery. All are ordinary features of science, recognizable by any practitioner, but hardly within the purvue of traditional philosophy of science. But consider just this last, the feeling of joy accompanying creation or discovery. This feeling seems naturally to accompany the bringing of a novel phenomenon materially into being in such a way that one nevertheless has confidence that one recognizes the phenomenon for what it is. One looks in vain for a place for or even a recognition of such an emotion in traditional accounts of the philosophy of science. In fact, philosophers of science such as Frege have gone so far as to oppose aesthetic feelings to those that accompany genuine scientific inquiry. But those who consider such emotions irrelevant to the process of experimentation and scientific inquiry have not fully grasped the kind of activity that science involves. Scientists frequently make a celebratory point of sharing a bottle of wine together after an epochal achievement. And when Brookhaven National Laboratory, in Long Island, New York, dedicated its first major particle accelerator, the Cosmotron, the first machine to accelerate particles past 1 GeV, the ceremony was quite a symposium. "At least one guest passed out on the table, and a Berkeley scientist set his tablecloth on fire. The final speaker, Dr. Detlev W. Bronk, president of Johns Hopkins University, mixed up the text of his speech with one he was scheduled to give in Canada, puzzling those still *compos mentis* with references to 'your King.' Because of the magnitude of the accomplishment, no one found the revelry excessive. 'A billion volts?' [accelerator physicist John] Blewett later remarked,' – that was one helluv'an achievement!'"[9] Is this merely an instance of overenthusiasm, of scientists getting carried away, that belongs only in popular literature about science – or, as the performance analogy suggests, is it a perfectly natural phenomenon of the psychodynamics of creativity about which there is an existing literature, and thus pointing to this link and to the literature about it, and thus *looking* for it, is not just a possibility but a duty of the philosopher of science? If a philosophy of science does not admit a role for such things within a comprehensive picture of science is it not legitimate to reject it as philosophy of *science*, and therefore as *philosophy*?

It is not a question of proving the analogy; as the Fermi example shows, argumentative analogies are *used* rather than *proven*. In philosophy, as in other disciplines aiming at the recognition of novel worldly structures, disclosure rather than proof is the culminating event. Against the background of the traditional philosophy of science, the analogy of experiment as performance is bound to seem imperfect and even incoherent as a description. But it is meant to be used argumentatively as a guide to phenomenological research into scientific experimentation – to aid in "filling in" a background. Only afterwards will there be enough background for it to seem descriptive. It is part of an *interpretation* of scientific activity, the development of a culturally and historically acquired understanding of it.

Scientific experimentation as performance

Those who have spent time at scientific laboratories and recording studios are bound to be struck by numerous similarities. Each place is a special environment in which every element – from materiel and architecture to personnel and facilities – has been tailored to fit the execution of the particular kind of action that happens there. To execute the action involves enormous preparatory work, in which a thousand tiny details affecting the outcome are supervised and checked. Each place requires a variety of kinds of participants, and is affected by a number of different economic, political factors, as well as by changes in direction of the respective fields (scientific research and music). Tensions can erupt between participants, some of whom may fend off accusations of being too worried about money or time, others of whom may accuse colleagues of overconcern with details rather than with the effect as a whole. To be sure, dissimilarities between lab and studio work are as profound as the similarities. But the similarities are useful as a tool for creating a perspective on experimental research far more adequate than that of confirming hypotheses or creating rhetorical claims.

The groundwork for applying the performance analogy is laid by the following: (1) Husserl's conception of the dual horizonality (internal and external) of perceptual phenomena, (2) the treatment of scientific entities as phenomena, via (3) the concept of readable technologies and an expanded (fully hermeneutical) conception of the external horizon, and accompanied by (4) a Deweyian-inspired notion of inquiry and the way it transforms inquirer and "world." The analogy itself requires connecting the three principal and related dimensions of performance – presentation, representation, and recognition – and the associated literature with features of science.

(A) *Presentation.* Performances are skillfully executed actions, or *presentations*, which aim at exhibiting profiles of phenomena. Scientific experiments are unique events in the world undertaken for the purpose of allowing something to become "visible" through readable technologies. What becomes visible is not something unique and peculiar to that event, but visible in different profiles via other types of performance. Scientific experiments can be executed well or poorly; what is seen stands out clearly in the former case and obscurely in the latter. Scientific performances are addressed to specific *prepared* communities and are responses to issues raised within those communities. Properly preparing and viewing the performances requires a detached attitude, one interested in seeing what is happening for its own sake rather than for some practical end. The outcome of the detached seeing of such performances, however, can be a deepened and enriched understanding of the world and our engagement with it.

One can distinguish between the *technology* and the *artistry of performance*. The *technology* of performance refers to passive forms of preparation – readable technologies, standardized props and procedures that one takes for granted. But the technology of an experiment is not independent of its end, the performance one wants to achieve. One has to decide the science one wants to look at, and then the performance conditions and signatures of that science, which may involve trade-offs; everything cannot be optimized simultaneously. A negotiation may thus take place between what one hopes to see, what is likely to be seen, and the technology available for such seeing. The *artistry* of experiment, on the other hand, refers to the active forms of preparation of a performance, and involves bringing a phenomenon into material presence in a way which requires more than passive forms of preparation, yet so that one nevertheless has confidence that one recognizes the phenomenon for what it is. For artistic objects "impose" themselves – they announce their presence as being completely or incompletely realized – but this imposition is not independent of the judgment and actions of the artist. The artistry of performance, too, is a hermeneutical process. One does not turn on the equipment and out pops a revolutionary new discovery. Generally, the presence of the novel or unexpected in an experiment perplexes an experimenter, who may respond by adjusting the equipment and introducing variations. The experimenter may still be confused, but with a better sense of what is confusing and what additional steps to take to alleviate the confusion. The anticipatory goal is the appearance of the phenomenon. But especially in the case of a genuinely novel phenomenon, one's anticipations of how it appears emerge together with the shaping of the experiment. One therefore "gives oneself over to" the performance the way one gives oneself over to a game or a role; one allows oneself to be surprised and transformed. One might refer to this as the *primacy of the performance*.

Like other kinds of performances, too, experiments are *holistic* in that they involve the simultaneous working together of a number of different kinds of elements. They are also *probative* (exploratory) in that what takes place is not fully predictable in advance. They are *provisory* in that they are always open to being redone. They are also *authoritative* in that they demand acknowledgment by those engaged in inquiry into the sort of activity being performed; there is always already a relatedness between the act and those witnessing the act. What appears in performance does so not magically, not as an exception to the order of the world like a magician pulling a rabbit out of a hat, but from an observed co-working of elements. One can see how it is done; nothing is concealed; it is the opposite of magic. However virtuosic or extraordinary, a performance is still a worldly event which is authoritative because its presence must be acknowledged in worldly terms. This is the case even when what looks or is asserted to be an appearing phenomenon turns out to be an epiphenomenon, for what then imposes itself is the demand to explain the epiphenomenality appearing in performance.

(B) *Representation*. Both the performance process and the performance product are structured by a *representation*. Every performance is related to some kind of representation, or program such as a text, score, or script.[10] This representation can then be used as a program for the performance of further experiments to explore that phenomenon. A program can itself become an object of knowledge; Maxwell's equations or Newtonian mechanics or Einsteinian relativity can be treated as a field of study in its own right, and found to contain discoveries, surprises, or inconsistencies. But the body of knowledge emerging from such study no longer concerns things in the world – phenomena – but a field of their possible profiles. Thus, theoretical knowledge is but one aspect of scientific knowledge, for science aims not only to probe possibilities, but also to show actualities.

A theory, then, *scripts* a phenomenon. But a script is related to a phenomenon in two ways. It structures the performance process (it "programs" the performance) and the performance product. A theatrical script, for instance, both structures the actions of the performers on the one hand, and describes what is heard (the play performed) on the other. Read noetically, a script is something to be performed; read noematically, it describes the object appearing in performance. In the context of experimentation, what is presented may show itself in such a way as to suggest another manner of scoring, so to speak. The act of experimental performance, while controlled by a representation, is done with the possibility in mind that what appears may do so in a way calling for a new representation. When a theory "works," expectations match what is disclosed, and what is disclosed fulfills expectations. These expectations are structured by the instrumentation, and the model helps us understand what that instrumentation is "reaching," or presenting to us. For the instrumentation does not present us with the model, but rather with some thing which through inquiry is seen to have (or lack) the structures of the model. The model of the double helix, say, allows us to move from profile to profile in the laboratory, instructing us in our manipulations and helping us to understand what is given through them.

In music, a successful score is one which, on the basis of standard practices, allows us to reenact the event. But what would be considered to be adequate scores (notational practices) in one period of history would not necessarily be considered to be adequate in another, when different kinds of music is played in different ways on different kinds of instruments. Scores, then, do not picture or represent ideal performances. Even assuming the truth of the fictitious story that

Mozart could not only recreate symphonies in his imagination upon reading a score but also *write* them in his head as well, he would have needed to rely on standardized practices (instrumental and notational) to do so. The *Noten im Kopf* to which he would be listening in the privacy of his internal theatre would be played on a clavichord, an instrument of his day, rather than on a pianoforte, an instrument perfected in the next century, or on a synthesizer, an instrument of our time. Scores or scripts are open-ended regarding possible performances. Theories thus cannot be viewed as picturing entities that exist apart from the life-world. A theory represents a phenomenon by scripting or scoring performances in which it appears. It represents an invariant which by standardized techniques and practices can be correlated with elements and operations of performances. A failed, inadequate, or incomplete representation is one which cannot be correlated with the elements and operations of a performance; it may indicate profiles that do not appear in performance, or may fail to be correlated with the entire set of what are evidently profiles of the same invariant. And theory-making alone cannot provide genuinely new information about phenomena, but only about possibilities for how already anticipated phenomena might appear; moreover, one still has to check whether the representing model provides a valid description of the represented phenomenon.

Theories in science *seem* different from representations of music, dance, and drama, because the former are constrained by (what is experienced in) the world. Although representations in these other art forms are also based on attention to experience, they nonetheless are viewed as the unique province of individual artistic imaginations addressing the social world, whereas scientific theories are thought to arise from a communal and objective encounter with the natural world. Had there been no Shakespeare or Beethoven, one may say, we would have no *Hamlet* or *Choral Symphony*; had there been no Maxwell we still would have something similar to the set of equations which bears his name and which have been used ever since their discovery. Maxwell's equations indeed have been used in widely different social, historical, and technological contexts for over a century now, and at first glance would seem to have the character of an immutable natural law. But our understanding of how these equations are *applied* to the situations at hand – hence, their function in a *theory* of electromagnetism – has changed with the context (which now includes, for instance, quantum mechanical situations). Furthermore, Maxwell's equations themselves have had a considerable history, and have changed markedly in how they are written down and in other respects.[11] Also, they also have problematic features which make it inevitable that in the future they will be further revised.

Thus what we need for an adequate account of theory is not only the Husserlian notion of invariance, or regularity of profiles under transformations, but also the Heideggerian notion of interpretation, for what fulfills or "counts as" a profile is a hermeneutical process and always developing. The assuring, enriching, and deepening of involvements with nature changes both the invariants able to be picked out as well as the profiles that count as fulfilling them. Discovery, say, of a spectral line in a predicted location might count as confirming the theory of quantum electrodynamics at one time – but at another, with more advanced instruments and more refined theoretical understanding, discovery of a slight discrepancy between the original prediction and the actual location might *also* count as confirmation.[12]

Here, too, there is a *primacy of performance*, for all theory is theory of performances of possible events. The development, reworking, and discarding of theories are all governed by how well they represent the invariances exhibited by sets of profiles in performance. Theories, to exaggerate somewhat to stress the point, are *fragile*. This is true in the dramatic arts, of course. Scripts and scores are not inviolate and sacred, but in actual use changed to fit what works in performance. To paraphrase Wilshire, theories describe the noetic-noematic structures of the encounters with human beings with nature via the mediation of instruments. These structures, and the theories that describe them, evolve over time, however slowly.

(C) Experiments are actions in which a suitably prepared community seeks *recognition* of phenomena. Recognition has a philosophical literature that reaches back to Aristotle. While it may seem odd to turn to Aristotle, of all people, and to his *Poetics*, of all places, for assistance in clarifying the paradoxes of the discovery process in science, this reference is made necessary by lack of appropriate treatment of the issue by traditional philosophy of science. For Aristotle, recognition is a transition from ignorance to knowledge, consists of a perceptual act, is the result of a concernful engagement with the world, and has unanticipated consequences. These four features form a basis for developing a suitable account of recognition for the context of science, elaborated with the aid of elements of other philosophical accounts: recognition concerns the *existence* of a phenomena or its presence in the world, is *dependent on the background context* or the perceptual abilities or state of readable technologies, is *temporal*, is *world-transforming*, and involves the *ever-present possiblity of misrecognition*.

In science, as other kinds of recognition, no rules can be formulated for infallibly recognizing novel phenomena, for recognition depends on an ever-changing background context and the relative novelty of what comes to be recognized. (What happens in the case of optical scanners and other instruments involving pattern recognition is that the background and types of foreground figures have been standardized.) When the background is not standardized and one wants advice on how to achieve a recognition it is difficult to improve upon the advice of Freud, one of the great attenders to performances, to pay "evenly hovering attention."[13]

But performance cannot be thought of in terms of its product alone; experiments as other kinds of performances must be *prepared* by an advance set of behaviors and decisions. It is therefore necessary to consider *production*, which refers to the set of decisions made in advance of a performance necessary for it to take place at all, and which makes it possible to speak of many of "the same" kind of performance. It would be a mistake, for instance, to think of a theatrical performance as exhausted in a video made of the opening night, which could then be replayed, rewound, and replayed again at will. A spectrum of production decisions and activities have to take place before that opening night in order to "shape" that opening night performance. If these decisions and activities were made differently, the outcome would differ as well. It is equally a mistake to think of experimentation as exhausted in the data that come from it. Experimentation is a process that reveals data, and the process takes time and is effected in phases. It involves its own set of decisions and activities made before the first experimental run, and how these decisions and activities are carried out affect the outcome of that run. The conditions of an experiment, like that of a theatrical production, do not have a single solution.

Advantages of the performance analogy

The vocabulary developed with the aid of the theatrical analogy and Dewey, Husserl, and Heidegger enables us not only to recognize and appreciate features of its practice whose significance is overlooked or dismissed in the traditional account, but also helps clarify: (A) traditional problems such as repeatability, referentiality, the role of mathematics, and others; (B) the relation between science as inquiry and science as cultural practice; and (C) certain issues regarding method.

(A) *Traditional Problems*. The *principle of the primacy of the performance* means that presentation, representation, and recognition are each in the service of the appearing of the phenomenon; data are *fluid*, theories *fragile*, and recognitions *stage-dependent*. For not all interpretations of the world are contextually legitimated; a performance can exceed the theory used to create and understand it, calling for new performances and theories. The unity of the sciences, and relations between different branches, are a function of the unity of and interrelations between the phenomena involved. A scientific phenomenon is something "behind" the data that *denominate* its presence and "behind" the theories by which it is represented. Data constrains theory through the mediacy of the phenomenon. Data describe how a phenomenon shows itself through readable technologies; they are relative to both the way the phenomenon shows itself and to the state of the technologies. Data, to exaggerate somewhat in order to emphasize the point against empiricists, are *fluid*, for they only come to mean something – and to legitimate itself as a set of data, as an appearance of a phenomenon – when it is incorporated with other sets of data that present other appearances into one picture of a phenomenon as appearing through different profiles.

Treating what is fundamental in science as *phenomena* helps to solve several traditional problems in the philosophy of science. One is the problem of *referentiality*, or what happens when theories change. In the light of the schema developed here, it is a matter of course that as the methods and techniques of science change, so does the appearance of phenomena and the theories we have to write for them. The two-fold horizonality of phenomena leads to a dynamical view of scientific entities; they are phenomena whose appearances are constituted by our praxes, whose appearances raise expectations about their other profiles, and these expectations are then refined or replaced depending on whether these profiles are fulfilled or not. Neither the changing appearances of scientific phenomena nor the changing theories by which they are described pose problems in describing their identity. In the event of a substantial shift in the outer horizon, when a phenomenon may appear markedly differently, it is still given as having had markedly different profiles in the previous outer horizon. This is part of the meaning of phenomena as historical. Another traditional problem is *repeatability*, for it is unclear what repetition of an experiment can mean. Each new scientific experiment differs from all others, involving different personnel, equipment, procedures, materials, computers, and so on, and even ongoing experiments often change as scientists try improvements to increase sensitivity or take advantage of new developments. How, then, can we call two different interventions into nature "the same?" But if scientific experiments constitute their objects as phenomena, one is not under any illusion that any two experiments, any more than any two acts of perception, can be identical, nor that they need be in order to have the same phenomenon appear. One can 'see' the same thing from different angles and even when it shows itself entirely differently. If, however, one agrees on the relevant background of equipment, practices, theories, and skills, it would make sense to say that one repeats an experiment. One aim of inquiry is to discover the invariant behind a series of profiles. This invariant does not appear through one, but the phenomenon may have to be transformed again and again befaor what is invariant shows itself adequately enough to be described. This, then, is reason for preoccupation with repetitions, rather than checking; they are not repetitions, but generation of multiple profile appearances. A third traditional problem has to do with the role of mathematics in science. Eugene Wigner once wrote an essay entitled, "The Unreasonable

Effectiveness of Mathematics in the Natural Sciences," which concludes with an admission of defeat in his attempt to comprehend this effectiveness: "The miracle of the appropriateness of the language of mathematics for the formulation of the laws of physics is a wonderful gift which we neither understand nor deserve."[14] But, as already mentioned, mathematics and invariants are not related as two things that magically mirror one another, but as process and product.

(B)*Science as Inquiry and Science as Cultural Practice.* Scientific knowledge is on the one hand a social product, the outcome of a concrete historical, cultural, and social context, and is therefore affected by the presence of political, economic, institutional, psychological, and other forces that are always at work in such contexts; on the other hand, scientific knowledge has a certain objectivity or independence from these contexts. Any appropriately structured inquiry into science at some point must clarify this *antinomic* character of scientific knowledge. The theatrical analogy shows a preliminary way of recognizing both the presence of a certain objectivity and the place of political, economic, institutional, and psychological factors, and of modeling their interaction. It would be as much of an error to *ignore* the presence of such factors as it would be to *reduce* the activity of science to them. Traditional philosophy of science has taken the former path, supposing that social, cultural, and historical influences either do not exist or that they are unimportant. Social constructivists and epistemological relativists, on the other hand, have chosen the latter path, considering socio-historical-cultural factors to be determinative. They underplay the constraints placed on the development of scientific experimentation and inquiry by the necessity to achieve the real presence of scientific phenomena in the laboratory as achieved through readable technologies. The theatrical analogy exposes the naïveté of both approaches. The unsophistication of someone who asserts that the theatrical world is exclusively about explorations of the Meaning of Human Existence is as evident as the cynicism of someone who claims that it is all politics or personalities. The point, of course, is that both views are right to an extent. The theatrical analogy, in conjunction with the schema of pragmatic hermeneutical phenomenology, allows us to *perceive* the situation aright, to *reconstitute* our perception of scientific experimentation, and to notice that scientific phenomena, like those of theatre, takes place amid a complex interaction of internal and external horizons. Scientific experimentation is simultaneously *ontological* and *praxical*, both concerned with the real presence and disclosure of invariants in the world on the one hand, and shaped by human cultural and historical forces on the other. The antinomic character of science gives rise to the temptation to overemphasize one of two different aspects; its objectivity (invariant structure) on the one hand, and its social construction on the other. Both temptations should be resisted, for performance involves the co-working of each.

The theatrical analogy allows us to create just such a model, in the difference between performance and production. While experimental *productions* always takes place within a social and historical context and thus bear the imprint of the various forces at work in that context (imprints which can be empirically studied), experimental *performances* reveal phenomena having a measure of independence from that context insofar as they reveal themselves as having profiles in other kinds of contexts. The concepts of performance and production thus allow a better understanding of the interrelation of science as inquiry and science as cultural practice.

(C) *Method*. Ernst Mayr has decried the so-called methodological distinction often drawn between experiment on the one hand, and comparison and observation on the other, and the utilization of such a distinction to mark an important difference in status between physical and biological research. Mayr ardently defends comparison and observation, and champions the validity of the conclusions that they yield.[15] But given the primacy of performance, this distinction becomes unimportant and Mayr's defensiveness unnecessary. What matters is recognition, and whether it transpires through the encounter with profiles produced in the laboratory, or through profiles that appear in the field is unimportant. This is relevant to the question of whether or not there is a priority of some of the sciences; whether, for instance, physics, chemistry, and biology are more "scientific" than, say, geology or ecology. But achieving a recognition – the name for the process in which something is apprehended as a profile of a phenomenon – may take place whether we can manipulate the phenomenon so that all of its profiles appear at will in a laboratory, or have only limited access to profiles in the field. I can recognize something as a desk by walking around it or by seeing it at a distance; I can know something is a mountain even if I have never climbed to the top or have only one limited view of it. Recognitions achieved by laboratory manipulations (as in physics, chemistry, and biology, say) are on an equal par scientifically with those achieved through observations made in the field (as in geology, paleontology, oceanography, and ecology). The phenomenon itself determines the method used to present, represent, and recognize it.

Conclusion

The study of scientific experimentation is of necessity open-ended. It is impossible to pin down an "essence" of experimentation the way that one can, for instance, pin down the essential features of a triangle, so that one can speak

84

from then on with confidence about its past and future forms. (One might even ponder the possibililty of a historical *telos* to experimentation.) Experimentation does not aim at a closed, finished structure, nor does it seek to reify or confirm a structure; instead, it seeks continued inquiry. Experimentation is world-building. Experimentation consists of variations and explorations of the involvements of historically constituted practices with nature under the influence of historically developed theories to disclose different kinds and appearances of phenomena. To think one could exhaustively describe the nature of experimentation with finality presumes one could imagine in advance all of these involvements, practices, theories, and phenomena – like presuming one can foresee all possible play performances.

The performance analogy suggests a revision in the images with which we understand the revelation of truth in nature. One of the earliest is in Psalm 19: "The heavens declare the glory of God; and the firmament sheweth his handywork." That image – the *epiphany of Nature* – emphasized the visibility of its truths; they are openly disclosed and publicly available to all who would look. Medieval thinkers such as Paracelsus, however, came to appreciate that special training was required to read that handywork, and those who studied nature and were thus able to reveal his will occupied a growing role as intermediaries between God and the human world.

Galileo's revision of the image depicted the truth of nature less as revealed than as encoded, in the "open book of heaven" which was "written in the language of mathematics." The *book of Nature* image emphasized that the truths of nature were not perspicuous and visible but concealed and difficult to obtain, and implied that searching them out had to be a task for a specially trained professional class of individuals. The image was liberating in Galileo's time in that it directed the attention of inquirers toward the search for invariants of a mathematical character able to be discovered through experimentation. But the image is no longer liberating for us.

Relativity and quantum mechanics have called into question the relation between theory and world in a way that challenges the validity of the image.

The picture of nature suggested by developing the analogy of experimentation as performance implies yet another image incorporating facets of the first two: the *play of Nature*. Play is here meant as an infinite, ceaseless activity exhibiting a myriad of forms in as many situations. Yet this play is not chaotic or random but governed by patterned, discoverable constraints. Thus the image retains the open, revealed character of the activity – its accessibility to perception – but at the same time its scriptedness. This play transpires for the most part without human participation, but human beings can and do participate through interventions in order to supplement their awareness of the various forms of play, and furnish them with material for script-writing. But as their perceptions of the play become developed, deepened, and enriched, the scripts change accordingly.

Use of the theatrical analogy is as a kind of experiment itself. It engages with the subject to sketch out the kinds of aspects we should inquire into and the kinds of outcomes that would satisfy us. It requires a certain training to apply fruitfully. It is not definitive, but always open to reevaluation and reexecution. Finally, as for any truly consequential experiment, if it raises more questions than it answers, so much the better.

References

Bernstein, J. (1968), *Elementary Particles and their Currents* W.H. Freeman, San Francisco.

Crease, R. and Mann, C. (1986), *The Second Creation: Makers of the Revolution in 20th Century Physics,* Macmillan, New York.

Fermi, E. (1971), *Collected Papers*, Volume 1, Amaldi, E. , Persico, E., Rasetti, F. and Segrè E., (eds.), University of Chicago Press, Chicago

Freud, S. (1959), "Recommendations for Physicians on the Psycho-Analytic Method of Treatment," in *Collected Papers* Vol. 2, Basic Books, New York

Fuller, S. (1989), *The Philosophy of Science and its Discontents*, Westview Press, Boulder, Colorado

Hacking, I. (1983), *Representing and Intervening*, Cambridge University Press, Cambridge.

Heelan, P. A. (1990), "Hermeneutical Philosophy and the History of Science," in Dahlstrom, D. (ed.), *Nature and Scientific Method: William A. Wallace Festschrift*, Catholic University of America Press, Washington, D.C

Heelan, P. A. (1989), "After Experiment: Realism and Research," *American Philosophical Quarterly* 26:4: 297-308.

Heelan, P. A. (1989a), "Husserl's Philosophy of Science," in Mohanty, J.N. and McKenna, W.R. (eds.), *Husserl's Phenomenology: A Textbook*, University Press of America and The Center for Advanced Research in Phenomenology, Washington, D.C., pp. 387-427.

Heelan, P. A. (1988), "Experiment and Theory: Constitution and Reality," *Journal of Philosophy* 85/9:515-524.

Heelan, P. A. (1988a), "Husserl, Hilbert, and the Critique of Galilean Science," in Sokolowski, R. (ed.), *Edmund Husserl and the Phenomenological Tradition*, Catholic University of America Press, Washington, D.C., pp. 157-173.

Heelan, P. A. (1987), "Husserl's Later Philosophy of Science," *Philosophy of Science* 54: 368-390.

Heelan, P. A. (1986), "Machine Perception," in Mitcham, C. and Huning, A. (eds.), *Philosophy and Technology II*, Reidel, Boston, pp. 131-156;

Heelan, P. A. (1983), *Space-Perception and the Philosophy of Science*, University of California Press, Berkeley.

Heelan, P. A. (1983a), "Natural Science and Being-in-the-World," *Man and World* 16: 207-219.

Heelan, P. A. (1983b), "Natural Science as a Hermeneutic of Instrumentation," *Philosophy of Science* 50: 181-204.

Heelan, P. A. (1983c), "Perception as a Hermeneutical Act," *Review of Metaphysics* 37: 61-75.

Mayr, E. (1982), *The Growth of Biological Systems*, Harvard University Press, Cambridge.

Vico, G. (1968), *The New Science*, Bergin, T.G. and Fisch, M.H. (trans.) Cornell University Press, Ithaca.

Wigner, E. (1979), *Symmetries and Reflections*, Ox Bow Press, Woodbridge, Connecticut.

Notes

1. See, for instance, Fuller. See Hacking, p. 1.

2. Heelan, Patrick A. (1990); (1989); (1989a); (1988); (1988a); (1987); (1986); (1983): (1983a); (1983b); (1983c).

3. Heelan, (1989).

4. "*Tentativo di una theoria dell-emissione dei raggi 'beta,'*" *Ricerca Scientifica* 4 (1933): 491-5.

5. Despite its genesis in analogy, the paper was, as Abraham Pais comments, "a major document full of novelty." Pais, *Inward Bound*, p. 17.

6. "Seldom," writes Franco Rassetti, "was a physical theory born in such definitive form." Fermi, p. 539.

7. Bernstein, p. vii.

8. "It is another property of the human mind that whenever men can form no idea of distant and unknown things, they judge them by what is familiar and at hand." Vico, p. 60.

9.   Crease, R. P. "The History of Brookhaven National Laboratory Part One," *Long Island Historical Journal* 3:2, p. 185.

10.   Some theatre consists of freely improvised performances in which actors, without much apparent forethought, spontaneously and individually apply skills to whatever features of the environment inspire them – remarks and gestures of passersby, noises, and the like. Here, each performance naturally differs from all others. In most theatre, however, the content of performances is codified to some extent in the form of a script – and even "pure" improvisation has its constraints, and usually consists of blocks of previously formed material loosely connected by short improvisational sections.

11.   It is unlikely, for instance, that a graduate student in physics would recognize these equations in the form in which Maxwell originally wrote them down. For an example of an historical treatment of Maxwell's equations, see Yang, C. N. "Maxwell's Equations, Vector Potential and Connections on Fiber Bundles," in *Proceedings of The Gibbs Symposium* (Yale University, May 15-17, 1989), pp. 1-2.

12.   Crease and Mann, Chapt. 7.

13.   Freud, p. 324.

14.   In Wigner, p. 237. See also Mark Steiner, "The Application of Mathematics to Natural Science," in *The Journal of Philosophy* 86:9 (Sept. 1989): 449-480, which cites some similar expressions of bafflement concerning the privileged role of mathematics in natural science, but which makes no more headway than Wigner in elucidating it.

15.   Mayr, pp. 30-2.

5 The Deconstruction of Some Paradoxes in Relativity, Quantum Theory, and Particle Physics

Simon V. Glynn

Introduction

Following the success of the physical sciences from the 16th Century onwards, many in the social and human sciences attempted to emulate their methods. Others, following Dilthey who argued that "Nature we Explain: Man we Understand," were of the opinion that the "objective" epistemology, and causal explanation, characteristic of the study of physical objects, was singularly inappropriate to the study of human subjects, and consequently they came to adopt hermeneutic and phenomenological, and latterly structuralist and deconstructionist approaches.

Moreover, recent developments in the natural sciences in general, and particle physics, quantum theory and relativity in particular, have given rise to problems and paradoxes which throw into question the Cartesian dualism and Lockean empiricism which have, broadly speaking, provided the epistemological foundations for the physical sciences. But while Husserl, Heidegger and Derrida etc. have raised general questions concerning many of the epistemological presuppositions of the, so-called objective, natural sciences, and Kuhn, Feyerabend and others have demonstrated the mechanisms of paradigmatic and methodological shifts within the natural sciences, what has not so far been recognized is that, as I shall try here to demonstrate, many of the problems and paradoxes associated with these recent developments in the natural sciences may be resolved or dissolved when approached from the epistemological perspective of phenomenology, hermeneutics, and their structuralist and deconstructionist offspring. Thus, so far from the human and social sciences adopting the epistemology and methodology of the natural sciences, on the contrary the natural sciences might do well to adopt the epistemology and methodology of the human and social sciences. Indeed we shall see that, while from the *Modern* perspective afforded by Cartesian or neo-Cartesian dualism and traditional forms of empiricism, complementarity, indeterminacy, field theory, holism, S-matrix theory, hadron bootstrap and many of the other formulations of quantum mechanics, particle physics and relativity theory, etc., seem to be inherently problematic and paradoxical, they may be unproblematically understood and unparadoxically formulated from the *postmodern* perspective.

> The reciprocal relationship of epistemology and science is of a noteworthy kind. They are dependent upon each other.
> – A.Einstein

I: The theory

Naïve empiricism and the phenomenological reduction

Husserl realized that many of the epistemological problems and paradoxes which confront us arise from unjustified assumptions or presuppositions. He therefore insisted that, in contrast to the "naive empiricism" of Locke and Co. – which has provided, broadly speaking, the epistemological foundations for much of what Thomas Kuhn has christened "normal science" – phenomenology restrict itself to giving *presuppositionless descriptions* of the realm of immediate experience or phenomena; of the phenomenologically *reduced* realm as it is sometimes called.[1]

Thus while traditional empiricism distinguishes between the so-called "object-as-it-appears" and the so-called "object-in-itself," and consequently accepts a dualistic division between "subjective" appearances and "objective" reality, Husserl points out that such a distinction is based precisely upon the fact that, in contrast to the "object-as-it-appears," the "object-in-itself" is NOT given in immediate experience. Consequently, the belief in "objects-in-themselves" is empirically unverifiable, a prejudice or presupposition that is therefore more properly a matter for metaphysical speculation, than for any empiricism worthy of the name.[2] Indeed, not only are "objects-in-themselves" inexperienceable in principle, but as Husserl goes on to demonstrate, they are, in addition, unnecessary in practice.

Suspending judgement on the existence or otherwise of such "objects-in-themselves," (practicing the phenomenological bracketing or *epoche*[3] as it is called), he shows that we are still able, on the basis of our immediately given experiences alone, to distinguish the experiencer from the experienced, appearances from that which appears, consciousness from its objects, ideal, or so-called "mental" objects, (eg. the fantasized unicorn, remembered home of my childhood, or dreamt unicorn), from each other and from physical objects (eg. the chair on which I am now sitting, or the desk in front of me).[4]

Therefore, Husserl concludes that we have as little use for "objects-in-themselves" as we have empirical evidence to support their hypothesized existence.

The hermeneutics of perception and the deconstruction of the "objective world of facts"

Now this ability to constitute objects "intentionally[5]," which is to say wholly and solely on the basis of experience, without invoking or appealing to a realm of "things-in-themselves" existing outside or beyond such experience, implies what Kant referred to as a "transcendental unity of apperception ... through which all the manifold given in an intuition, (or experience as we would now refer to it), is united in a concept of the object."[6] In other words, given that Husserl has bracketed or suspended judgement on the existence of things-in-themselves, then they can no longer be taken to account for the connections displayed between our experiences, on account of which we take *many* experiences to be experiences of *one* and the self-same object. Therefore, the fact that, as we have just seen, our experiences or perceptions nevertheless continue to remain "combined" into "objects" may be taken to imply that we ourselves are actively responsible for this unifying synthesis; that our *perception* is necessarily mediated by our *conceptions*, or, to put the same point otherwise, our *descriptions* of the "objective" world, along with this world itself, are already the product of our *interpretive preconceptions*.

Thus while phenomenology may indeed be able to avoid some of the presuppositions and preconceptions which have plagued naïve empiricism, it is certainly not able to avoid all such preconceptions. As Merleau-Ponty affirms, "The most important lesson which the reduction teaches us is the impossibility of a complete reduction,"[7] for, as he affirms, "...perception is just that act which creates at a stroke, along with the cluster of data, the meaning which unites them,"[8] the concepts which make possible the unifying synthesis of which we have spoken.

This view is empirically attested to by the cognitive psychology of the *Gestaltists* and Ames and his school, who have demonstrated that even our most basic perceptions of the world are mediated by the, often implicit or unconsciously held, concepts, by which we attempt to render what *might* otherwise be a "...boomin', buzzin', confusion," coherent and predictable.[9] Indeed even philosophers such as Quine for instance, insist that we regard "... physical objects ... as ... device(s) for working a manageable structure into the flux of experience."[10] As Einstein affirms:

> ...the formation of the word, and hence the concept "ball" (for example) is a kind of thought economy enabling the child to combine very complicated sense impressions, (or experiences as we would prefer to say), in a simple way ... Mach also thinks that the formulation of scientific theories ... takes place in a similar way. We try to order the phenomena to reduce them to simple form, until we can describe what may be a large number of them with the aid of a few simple concepts.[11]

Thus while science and scientific *theories* are grounded upon the *"facts,"* upon the *perception* of objects and their relations, insofar as these *perceptions*, and the "facts" revealed in them, are the constituted products of our *conceptual interpretations*, then science and its theories are grounded upon such conceptually mediated perceptions, and may be seen as nothing other than a reflective, conscious, extension of this process of rendering our experiences coherent and therefore predictable. In view of this it comes as no surprise to learn that hermeneutic philosophers, such as Heidegger, believe that our perceptions of scientific "facts" of the world are at least as likely to reflect our theoretical conceptions as *vice versa*:

> The greatness and superiority of natural science during the sixteenth and seventeenth centuries, (Heidegger tells us), rests in the fact that all the scientists were philosophers. They understood that there are no mere facts but that a fact is only what it is in the light of the fundamental conception...[12]

Indeed, even those philosophers of science normally regarded as hostile to the hermeneutic tradition, such as Karl Popper for example, have had to concede that "...observation is always *observation in the light of theories*."[13] Clearly then, what Nietzsche called "The Myth of Immaculate Perception"[14] is dispelled forever, and even Husserl comes, in his later work, to recognize:

> ...the *naivete* of speaking about "objectivity" (and of the scientist) who is blind to the fact that ... the objective world itself ...(the everyday world of experience as well as the higher level conceptual world of knowledge) are his own *life-constructs* developed within himself...[15]

Thus we can see that even, (or perhaps especially), the supposedly "objective" world, be it the world given to us in immediate experience, or the scientific world, can be decontructed; can be shown to be the hermeneutically constituted reflection of our own theoretical interpretations or conceptual perspectives.

Deconstructive coherence

This does not mean however that there are no constraints upon such deconstruction, that any theories and theoretical conceptions are as acceptable as any others. For while perceptions may be ambiguously dependent upon implicit conceptions, if scientific theories are to explain and / or predict the behavior of (the objects and events given in) such perceptions, there must be some accommodation between these explicit *theories* and the implicit theoretical *conceptions* structuring the perceptions which they seek to explain and / or predict; they must be coherent with one another. While if, as we have argued, the pre-reflective concepts that mediate our most basic perceptions, are, like the reflective concepts or theories of science, also attempts to render our experiences comprehensible, then they too, if they are to be successful in this regard, must also be coherent with one another.

Consequently, as Lakatos insists: "It is not that we propose a theory and Nature shouts *NO*; rather we (implicitly and explicitly) propose a maze of theories and Nature may shout *INCONSISTENT*."[16] Thus, as Paul Ricoeur has pointed out:

> ...the hermeneutic circle ... (as this circle of interpretation is known) is not a vicious one ... (for) to the procedures of validation there also belong procedures of invalidation similar to the criteria of falsifiability proposed by Karl Popper ... here the role of falsification is played by the conflict between competing interpretations[17]

But whether potential incoherences are resolved by adjusting our implicit *preconceptions* or *conceptions*, or our fully reflective scientific *theories*, or both, remains a question of choice; accepting of course, that unless one wishes to embark upon a complete paradigm shift, theoretical conceptions at both levels should remain consistent with our other, already accepted, theoretical conceptions, ideologies, etc.

Thus in this way then the ideology of a *culture* enters into its view of *nature*, which is therefore subject to *deconstruction*, while despite the consequent absence of a theory independent, ideologically unsullied, privileged realm of facts "in themselves" to constrain them, interpretations nevertheless remain constrained by the requirement that they be coherent. However, here we should note that the need for coherence issues not from some purely formal, transcendental or metaphysical dictate, but, as we shall examine in greater detail presently, from a pragmatic interest in rendering experience consistent or comprehensible so that we may have the power to predict, explain and ultimately manipulate it, a point to which we shall return *via* some preliminary reflections upon "the subject."

Phenomenology, hermeneutics and existentialism, and the deconstruction of the subject

If, as we saw previously, we have no experience of, and therefore no empirical evidence for the existence of, a world of "objects-in-themselves" existing beyond experience, it should be equally clear that the same can be said of a "subject-in-itself." Indeed despite what some have seen as a certain equivocation concerning the status of the Transcendental Ego in Husserl's philosophy, so far at least as the empirical ego, or experiencing subject, is concerned, he is insistent that, "...the experiencing ego is still nothing that might be taken *for itself* and made into an object of enquiry on its *own* account. Apart from its "ways of being related" or "ways of behaving" it is completely empty of essential components."[18]

That is to say, that as Sartre recognized, "...consciousness is not for itself its own object...,"[19] not an object of consciousness, but the subject, which has no existence apart from the capacities, – eg. perceiving, remembering, feeling, valuing, judging, initiating and guiding actions, etc. – which reifying reflection[20] usually takes it to account for. Thus recognizing that we have no experience, sensory or non-sensory, of an egological subject as such, Sartre insists that "...the transcendent I," by which he means all egological conceptions of consciousness as a thing existing independently or separately from these capacities, "must fall before the stoke of the phenomenological reduction."[21] Indeed, Sartre insists that "All the results of phenomenology begin to crumble if the I is not, by the same title as the world, a relative existent."[22]

Thus we can see that the notion of an independently existing subject, no less than that of an independently existing world of objects, has been deconstucted in favor of an empirical relationship in which experiencer (or "subject") and experienced (or "object") are reflectively distinguishable, yet ultimately inseparable "poles" of experience.[23]
As Henry Margeneau sums up:

> It is wholly unwarranted to start a theory of knowledge with the ontological premise characterizing the spectator spectacle distinction. If experience on proper analysis, invests this distinction with meaning, we will be ready to accept it, but even then only as a property of the content of experience, actual or possible.[24]

Heidegger also concurs, insisting that "...a bare subject without a world never 'is' proximally, nor is ever given."[25]

But he further points out that the cognitive reflection which will come to identify the "cognizer" and the "cognized" as poles of cognition takes place on the basis of a pre-cognitive existential or ontological relation to the world[26] which cognitive or epistemological reflection *dis-covers* as always already existing. That is to say that, prior to theoretical reflection human *Dasein*, or "Being-in-the-world" takes the form of a practical (*Zuhanden* or "ready-to-hand") relationship.[27] In which case, as Michel Foucault has shown, the human "subject" may be deconstructed in terms of the web of economic, political, social, cultural, sexual and other power relations, which constitute this practical relationship, from which it subsequently emerges.[28]

Thus it would seem that here again, the roots of deconstruction, this time of the "subject," are to be found in phenomenology, existential phenomenology, and hermeneutics.

Pragmatic deconstruction of the metaphysics of truth

And if *Dasein's* relationship to the world is, in the first instance at least, practical, then so far from being those mythically rational, disinterested spectators that the reifying reflection[29] of Enlightenment though would make of us, our relationship to the world is, on the contrary, characterized by Care (*Sorge*) and Concern (*Besorge*).[30] This being so then even, or perhaps especially, the "objective," "value-free," quantitative world as it supposedly appears to passive disinterested reflection, is in fact the product of abstraction from the world as it is initially given to us, as permeated with qualitative significance and value,[31] an active abstraction by a subject very much interested in power and mastery over things.[32] As Habermas, anticipating a point of which the postmodernists have made much, puts it, "... objectivism ... conceals the connection of knowledge and interest."[33] Indeed, ".. *knowledge and interest are one.*"[34] Thus as Nietzsche explains:

> The compulsion to form concepts, genera, forms, ends and laws ("*one world of identical causes*") should not be understood as though we were capable through them of ascertaining the true world, but rather as the compulsion to adopt to ourselves a world in which our existence is made possible. Thereby we create a world that is calculable, simplified, understandable, etc...[35]

Clearly then the metaphysics of truth has been deconstructed in the name of what Nietzsche, with perhaps too little regard for the implications of such a phrase, has referred to as a "useful falsification,"[36] in favor, that is, of humble pragmatism originating from our primordial practical participation in the world. Thus modernity is unable to exclude the subject from the "world picture," is unable to replace the existentially interested subject participating in the world, with a disinterested outside observer, the spectatorial subject of Cartesian reifying reflection. Indeed as Heisenberg, for example, sums up:

> When we speak of nature in the exact science of our age we do not mean a picture of nature so much as a *picture of our relationship with nature*. The old division of the world into objective processes in space and time, and the mind in which these processes are mirrored – in other words the Cartesian distinction between *res cogitans* and *res extensa* – is no longer a valid starting point for the understanding of modern natural science. Science, we find, is now focused on the network of relationships between man and nature; on the framework which makes us as living beings dependent parts of nature, and which we as human beings have simultaneously made the object of our thoughts and actions. Science no longer confronts nature as an objective observer, but sees itself as an actor in this interplay between man and nature.[37]

And it is my contention that phenomenology, hermeneutics and their structuralist and deconstructionist offspring provide us with an enormously helpful framework from within which to approach such a postmodern world view in general, and post-Newtonian physics in particular.

II: The practice

Relativity & the subject/object relation

Just as phenomenology denies empirical knowledge of anything existing in and of itself, beyond or outside experience, Einsteinian relativity denies any knowledge of a space and time existing independently of the observer. In accordance with Husserl's general thesis that all experiences of objects and their relations are "intentional," which is to say experiences of an experiencing subject, to whom, – as the hermeneuticists and poststructuralists, building on Husserl's notion of the Lifeworld (*Lebenswelt*) have argued, – they are therefore relative,[38] for "modern science" too, as Sartre affirms, "Man and the world are relative beings, and the principle of their being is the relation."[39] Thus Einstein specifically insists that the experience of space / time is relative to the observer. For example Bohr tells us:

In the past, the statement that two events are simultaneous was considered an objective assertion, one that could be communicated quite simply and that was open to verification by an observer. Today we know that 'simultaneity' contains a subjective element, inasmuch as that two events that appear simultaneous to an observer at rest are not necessarily simultaneous to an observer in motion. However, the relativistic description is also objective inasmuch as every observer can deduce by calculation what the other observer will perceive or has perceived ... but it is no longer possible to make predictions without reference to the observer or the means of observation. To that extent every physical process may be said to have objective and subjective features.[40]

Thus like phenomenology, relativity theory clearly recognizes that all empirical reports are relative to an observer, while in insisting that the sequence in which an observer will experience a given event will be absolutely dependent upon the observer's position, velocity and direction of motion *vis-à-vis* the event, Einstein is eschewing the disembodied Cartesian *spectator* of naive empiricism in favor of something more akin to the *participatory* subject of more recent Continental philosophy. Furthermore the tension between the relativity of our observations, and the absolute determinability of the sequence in which events will be observed by a particular observer, echoes the tension between "life-world" relativism and essentialism[41] reflected in Husserl's insistence that experience has essential features that are irreducible to the subjective flow of mere appearances.[42]

Existence as experienceable properties

Turning to atomic physics, Bertrand Russell, recognizing (and on this phenomenalism and phenomenology are in agreement) that we have no direct experience of atoms *per se*, has observed that: "For aught we know an atom consists entirely of the radiations that come out of it...the idea that there is a little hard lump there which is the electron or proton, is an illegitimate intrusion of common sense notions."[43] Here then we can see Russell, clearly and with good reason unhappy that a science claiming to be empirical should engage in such speculation, arguing that we should suspend judgement, or practice what is in effect the phenomenological reduction or "*epoche*" regarding that common sense belief in an unexperienced atom existing independently of its experienceable properties.

Nor is it any good to object that atomic and sub-atomic particles may be experienced with the aid of electron microscopes, cloud chambers, bubble chambers or other such devices, for all that is directly experienced in such cases

97

are interferences in interactive fields or tracks across chambers. Thus, the recognition of such interferences and tracks as evidence of atomic and sub-atomic particles owes everything to what Heidegger refers to as the "fundamental conceptions"[44] or hermeneutic preconceptions of the person interpreting such tracks, and nothing at all to the directly experienced facts. As Heisenberg affirms "...our experience of atoms can only be indirect: atoms are not things."[45]

Thus we can see that atomic and sub-atomic physics practices what can be described as a phenomenological *"reduction," of the object of inquiry to it's experienceable properties*, the concomitant of any thoroughgoing or radical empiricism, while it is this "reduction" which is the very precondition and justification of quantum physics, as we shall now see.

Properties as relations: the subject/object relation

In general then it should be clear that, empirically speaking at least

> ...the properties of any atomic object can only be understood in terms of the object's interaction with the observer. This means that the classical idea of an objective description of nature is no longer valid. The Cartesian partition between the I and the world, between the observer and the observed cannot be made when dealing with atomic matter."[46]

Indeed nowhere is this more clearly apparent than in Heisenberg's indeterminacy or uncertainty principle. Thus if we represent the probability of finding a particular particle at a given point with a given momentum by a "probability wave," in which the amplitude is positively related to the probability of it's being located at any given point, and the wavelength is inversely related to the particle's momentum, then it follows directly from the property of waves (i.e., that the shorter the wave packet the greater the spread in wavelength), that the more certain we are of the location of the particle the less certain we will be of it's momentum and *vice versa*. The process of measuring consequently influences the nature of the measurement. As Sartre insists, "Heisenberg's principle of indeterminacy ... Instead of being a pure connection between things ... includes within itself the original relation of man to things..."[47]

We are then no longer dealing either with passive spectatorial cognition, or with cognized objects existing "in themselves" independent of the cognizer and the process of cognition. On the contrary, we are dealing with a cognizing in which the cognizer is related to and interacts, through the process of cognition, with what is being cognized; in which cognizing subject and cognized object have been deconstructed in terms of a primordial relation. "...in the most recent phase of atomic physics," Heidegger confirms, "even *the object vanishes* ... the subject-object relation as pure relation thus takes precedence over the object and the subject."[48] In this then, modern science is wholly in accord with the tendency to deconstruct the dualistic conception of "subject" and "object" in favor of a unitary or holistic relation. As Wheeler concludes:

> Nothing is more important about the Quantum principle than this, that it destroys the concept of the world's "sitting out there"...one has to cross out the old word 'observer' and put in it's place the new word 'Participator'. In some strange sense the universe is a participatory universe[49]

Properties as relations: the object/object relation

Now just as we have seen above that the properties of objects are inextricably dependent on their relations to an observer, so too they are also inextricably dependent on their relation to each other. For example my understanding of the properties of water, (e.g., it's refractive index, it's density, etc.), is bound up with my understanding of it's relations to other objects, (eg. photons, wood and metal etc.), with which it interacts, and to which in turn the same considerations apply. Thus while I understand *water* as floating *wood* and dousing *fire*, I understand *wood* as being burned by *fire* and floating in *water*, and *fire* as burning *wood* and being doused by *water*. In short then, just as I can come to understand the meaning of a word by looking it up in a dictionary, which is to say by defining it in terms of its relations to other words, to which, in turn, the same applies, – the *semantic* meaning of a word deriving, as de Saussure has pointed out, from its *syntactical* relations to other words,[50] – so, as I have just now shown, *I can come to understand the empirical properties of an object in terms of its relations to other objects, and of course to the observer.* Concomitantly, the empirical *or experienceable properties which constitute an object are* thus reducible to, can be deconstructed in terms of, its *formal* or *structural* relations to other objects, to which, in turn, the same applies. Bohm explains that similarly also: "...at the quantum level of accuracy, an object does not have any 'intrinsic' qualities belonging to itself alone; instead it shares all it's properties mutually and indivisibly with the systems with which it interacts." Furthermore he continues: "... because a given 'object'... interacts at different times with different systems that bring out different properties, it undergoes continual transformation between various forms in which it can manifest itself..."[51]

Here then we have the foundations of a, perhaps neo-Leibnizian, holistic epistemology, in which the experienceable properties of an object, which is to say its empirical identity, is dependent upon the holistic system of relationships in which it is implicated; *formal relations* thereby constituting or generating the *existential content* in accordance with the claims of the structuralists. And because such "an object" can interact with different systems of relations, then – like the meaning of a word, which can never be definitively fixed, but can change or *differ* endlessly depending upon the context in which it appears, – the identity of such "an object" will never be fixed, but must be *deferred*, as it is endlessly constructed, deconstructed and reconstructed in terms of the relations into which it enters.

Existence as a relation

On the basis of what, for reasons of space, must here remain necessarily impressionistic, generalized and tentative epistemological insights, we may now perhaps be able to dissolve, or "deconstruct" some of the apparent paradoxes confronting those at the forefront of modern physics. Let us take one of the best known of such paradoxes, Niels Bohr's theory of complementarity, as a test case.

According to observations, light displays the properties both of a wave (e.g., the double slit experiment) and a particle (e.g., the photoelectric effect), and is consequently referred to as a "wavicle." Thus, like the *meaning* of a word, – for example "volume," which may change depending upon whether I ask you to "turn up the volume," "pass me that volume," or "estimate the volume of water in my radiator," – it would seem that, in accordance with David Bohm's general insight (above), the *properties* of light may undergo *Gestalt* like transformations between wave and particle (Bohr's wavicle) as we interpret it firstly from the perspective of its interactions with or relations to one system and then another.

But it can certainly be objected that while words may indeed be polysemic, and that similarly the properties which objects display may be context dependent, nevertheless, no matter how we may vary the context, no *one* thing can be both a wave and a particle as this contradicts the law of identity. However, if, alternatively, we abandon naive empiricism in favor of radical empiricism (or phenomenology), which reduces the very *existence* of the object to it's experienceable *properties*, properties which, as we have just seen, are in turn inseparable from the structure of their *relations* to other "objects," then it consequently becomes evident that the very *existence* of the object arises out of, and therefore may be reduced to or "deconstructed" in favor of, the system of *relations* in which "it" is implicated. In such circumstances a "wavicle" for instance, far from being understood as a "thing-in-itself" with, what are in consequence, logically contradictory properties, would be reconceived as simply the reified product of two complementary systems of relations or interactions; such properties, no longer being ascribed to a single "thing," no longer seeming contradictory. Thus such a deconstruction clearly solves, or more properly dissolves, this otherwise insoluble paradox.

Affirming this insight Heisenberg, in absolute accord with the phenomenological reduction, – the insistence that we have no direct empirical knowledge of objects existing independently of our experience,– tells us that: (with the advent of) "The concept of complementarity ... the idea of material objects that are completely independent of the manner in which we observe them proved to be nothing but an abstract extrapolation, something that had no counterpart in nature."[52]

Nor should this surprise us, according exactly as it does with our earlier recognition that such "objects" may be regarded, like "theories," as unifying structures which render experiences coherent.

Therefore, we can now see that by *(1) adopting a phenomenological epistemology and consequently suspending judgement on whether or not there is any empirically inexperienceable substratum of "things-in-themselves" existing beyond or behind and causing such experiences, and by (2) accepting our earlier hermeneutic insight that "objects" are in fact nothing other than interpretive theories,* we are able to deconstruct the "wavicle" as two complementary systems of interpretation which, *qua* interpretations, make no claims regarding irreducible "things-in-themselves" whatever, and consequently involve no paradoxes or contradictions. This of course in absolute contrast to those, such as naive empiricists and dualists, who, having failed to complete the phenomenological reduction, cannot help but regard the "wavicle" as an independently existing "thing-in-itself," whose properties are consequently self-contradictory.

As this test case clearly demonstrates beyond the shadow of a doubt, phenomenology, hermeneutics, structuralism and deconstruction do indeed enable us to resolve such paradoxes. And although considerations of space make it impossible to demonstrate here, it should be equally clear that the application of the same epistemology to anti-matter, virtual particles, positrons, and a range of other seemingly paradoxical "entities" of modern physics might reasonably be expected to produce similar results. In other words, once we cease reifying such "entities," and come to regard them, not as "things-in-themselves," but rather recognize them for what they are – analogues, or what Heidegger has referred to as "Nature... mathematically projected,"[53] – as part of the "garb of ideas,"[54] as Husserl calls the abstractly extrapolated theories by which we endeavour to interpret our experiences, then the paradoxes seemingly inherent in these conceptions dissolve just as they did in the "wavicle" test case above.

Further, not only may this approach be expected to work in resolving paradoxes concerning *entities*, but it may equally be expected to work, with appropriate modification, in resolving those paradoxes associated with inherently *relational conceptions* such as quantum jumps, event horizons, N-dimensional and curved space for example. And it may be expected to do so by drawing our attention to the fact that such formulations also are no more nor less than theoretical concepts by which we endeavour to render our experiences coherent, and should therefore not be reified; an insight which again enables us to solve, or more accurately dissolve, the paradoxes they otherwise apparently manifest.

Holism

To return to our main theme. Having phenomenologically reduced the existence of the "*things*" of particle physics to their *experienceable properties*, we have seen that these in turn are reducible to their *relations* to other "things" or particles to which, in turn, the same applies. Thus we are led *via* our initial phenomenological insight to the structuralist conclusion of quantum field theory that, as Stapp tells us, "An elementary particle is not an independently existing unanalyzable entity. It is in essence a set of relationships which reach out to other things,"[55] to which, following the same reasoning, exactly similar considerations apply.

On the other hand, however, according to this same theory, all interactions take place through the exchange of particles.[56] Therefore while particles may be defined in terms of their interactive relations with other particles, these interactive relations may in turn be defined in terms of the exchange of particles.

This so called S-matrix theoretical view of the relation between particles and their interactive relations is then a classical *Gestalt* background / foreground relationship in which what gets interpreted as a particle depends upon what gets interpreted as an interaction and *vice versa*; a view clearly paradoxical, even incomprehensible, to anyone who has not yet made the deconstructionist move and still insists upon regarding objects, including particles, as independently existing "things-in-themselves."

As we can now clearly see, in insisting that all so called "particles" are reducible to experienceable properties, and that all such properties depend, in turn, upon interactive relations, quantum physics, like phenomenology, is rejecting all attempts to explain such experiences in terms of an independently existing, empirically inaccessible, substance, preferring, like structuralism, to explain such experiences in terms of interactive relations. Therefore, while an "electron," for example, may be regarded, as Bohm suggests, as a particle interacting with different systems, equally we may make the structuralist move of reducing content to form and regarding it instead as a wave of interactions within a field of interactive relations. As Weyl explains:

> ...a material particle such as an electron is merely a small domain of the electric field within which the field assumes enormously high values ... such an energy knot ... propagates through empty space like a water wave across

the surface of a lake; there is no such thing as one and the same substance of which the electron consists all the time.[57]

In light of what we have seen we come therefore to recognize that for quantum physics, as for phenomenology, there is no independently existing substratum or substance, but that, as with structuralism, it is a web or network of interactive relationships which constitute "entities" as we experience them; that "each hadron (small particle) is held together by forces associated with the exchange of other hadrons ... each of which is in turn held together by forces to which the first hadron makes a contribution."[58] This then leads ultimately to the deconstructionist epistemology of hadron "bootstrap theory," according to which, "...each particle helps to generate other particles, which in turn generate it."[59]

Bohm and Hiley elaborate:

> One is led to a new notion of unbroken wholeness which denies the classical idea of analyzability of the world into separate and independently existing parts ... we have reversed the usual classical notion that independent 'elementary parts' of the whole are the fundamental reality and that the various systems are merely the particular contingent forms and arrangements of these parts. Rather we say that the inseparable quantum interconnectedness of the whole universe is the fundamental reality, and that relatively independently behaving parts are merely particular and contingent forms within the whole.[60]

Such a holistic universe, as we have seen, clearly includes the subject; a subject which, as we saw previously, is itself reducible to its relations to other "objects" and subjects, etc. The ontology of "things" having thus been replaced by that of "relations," we now confront a structuralist conception of the universe in which *content* is generated out of, or by, *form*, and can therefore be constructed and deconstructed in terms of such formal systems of relation. This then is the structuralist, and ultimately deconstructive, concomitant of the phenomenological reduction.

Summary

Starting with the hermeneutic recognition that all *descriptions* are always already *interpretations*, we saw how it followed directly that objects could be regarded, like theories, as devices by which we endeavor to render our experiences harmonious and coherent. Following the "radical empiricism" of phenomenology, we refrained from reifying such theoretical concepts or constructions, from gracing them with the status of independently existing objects. In so doing we reversed a trend exemplified by the founders of early modern science and philosophy such as Galileo, Bacon, Newton, Locke and Co., who not only legitimately *distinguished* the existence of objects from the empirical properties they displayed in their relations to an experiencing subject, but illegitimately sought to *separate* them. Recognizing, along with the structuralists, that such properties are, in addition, inseparable from an object's relations to other objects, we concluded, as does postmodern ontology, that each object, although analytically *distinguishable* from other objects, remains existentially *inseparable* from them.

Accordingly, we sought, along with earlier, supposedly "primitive," thinkers, to unite the *existence* of objects with the *properties* or "powers" they displayed in their *relations*. Recognizing, however, that, phenomenologically or "radically empirically" speaking, objects are nothing "in-themselves" apart from the properties or powers they display in such *relations*, we, (in contrast to these "primitive" thinkers, who attributed such properties or powers to objects), attribute the existence of objects to these properties or powers, and thus ultimately, following the structuralists, to the relations in which such properties or powers are manifested. And in this, as we have demonstrated throughout, we are in accord with contemporary science. As Habermas affirms, "...with the origin of modern empirical science the classical metaphysical concepts of substance have been replaced by the concepts of relation."[61]

As we can now finally see, then, notwithstanding many scientists' failure to recognize the fact, the epistemology /ontology of the "new physics" in particular, and much of contemporary science in general, is in reality far closer to phenomenology, hermeneutics, structuralism and poststructuralism, and to the postmodern, than to the Lockean empiricism and Cartesian dualism of modernity. Consequently, rather than reverting to an *ad hoc* defence of the "foundational realism" associated with modernity's neo-Lockean and Cartesian epistemology, contemporary science, when faced with apparent epistemological contradictions, would clearly do better to look to the postmodern to dissolve such contradictions and the paradoxes associated therewith.

References

Babich, B. E. (1994), *Nietzsche's Philosophy of Science: Reflecting Science on the Ground of Art and Life*, State University of New York Press, Albany.
Bohm, D. (1958), *Quantum Theory*, Prentice Hall, Englewood Cliffs, NJ.
Capra, F. (1976), *The Tao of Physics*, Fontana, London.

de Saussure, F. (1959), *Course in General Linguistics*, Philosophical Library, New York.

Foucault, M. (1970), *The Order of Things*, Pantheon Books, New York.

Habermas, J. (1978), *Knowledge and Human Interest*, Shapiro, J. (trans.), Heinemann, London.

Heidegger, M. (1978), "Modern Science, Metaphysics & Mathematics" in Krell, D. (ed.), *Martin Heidegger:Basic Writings*, Routledge & Kegan Paul, London.

Heidegger M. (1978a), "The Age of the World Picture," in Krell, D., *Martin Heidegger: Basic Writings*.

Heidegger, M. (1977), "The Question Concerning Technology," in *The Question Concerning Technology and Other Essays*, Lovitt, W. (trans.), Harper & Row, New York.

Heidegger, M. (1962), *Being & Time*, Maquarrie J. & Robinson, E. (trans.), Harper & Row, New York.

Heisenberg, W. (1971), *Physics and Beyond*, Pomerans, A.J. (trans.), Anshen, R.N. (ed.) Harper & Row, New York.

Heisenberg, W. (1958), *The Physicist's Conception of Nature*, Hutchinson, London.

Husserl, E. (1970), *The Idea of Phenomenology*, Alston, W. & Nakhnikian G. (trans.) , Nijhoff, The Hague.

Husserl, E. (1970a), *The Crisis of European Science and Transcendental Phenomenology*, Northwestern University Press, Evanston.

Husserl, E. (1969), *Cartesian Meditations*, Cairns, D. (trans.), Martinus Nijhoff, The Hague.

Husserl, E. (1962), *The Ideas*, Boyce-Gibson, W.R. (trans.), Collier-Macmillan, New York.

Kant, I. (1958), *The Critique of Pure Reason*, 2nd ed., Kemp-Smith, N. (trans.), Macmillan, New York.

Lakatos, I., (1976), "Falsification & the Methodology of Scientific Research Programmes" in *Criticism & the Growth of Knowledge*, Cambridge University Press, London.

Mehra, J. (1973), *The Physicist's Conception of Nature*, Riedel, Dordrecht.

Margeneau, H. (1950), *The Nature of Physical Reality*, McGraw-Hill, Maidenhead.

Merleau-Ponty, M. (1962), *The Phenomenology of Perception*, Smith, C. (trans.), Routledge and Kegan Paul, London.

Nietzsche, F. (1954), *Werke*, 2nd ed., Schlechta, K. (ed.), Hanser, Munich.

Popper, K. (1980), *The Logic of Scientific Discovery*, Hutchinson, London.

Quine, W.V.O.(1953), *From a Logical Point of View*, Harvard University Press, Cambridge MA.

Ricoeur, P. (1976), *Interpretation Theory: Discourse and the Surplus of Meaning*, Texas Christian University Press, Fort Worth, Texas.
Ricoeur, P. (1974) , *The Conflict of Interpretations*, Ihde, D. (ed.), Northwestern University Press, Evanston.
Russell, B. (1927) *The Outline of Philosophy*, Allen Unwin, London.
Sartre, J.-.P. (Undated), *The Transcendence of the Ego*, Williams, F. & Kirkpatrick, R. (trans.), Farrar, Strauss & Giroux, New York.
Sartre, J.-P. (1956), *Being & Nothingness*, Barnes, H. (trans.), Philosophical Library, New York.
Sartre J.-P. (1947), *The Emotions: Outline of a Theory*, New York Philosophical Library, New York.
Weyl, H. (1949), *The Philosophy of Mathematics and Natural Science*, Princeton University Press, Princeton, NJ.

Notes

1.   For a full discussion of the Phenomenological Reduction see Husserl, (1970), Lectures II-III & Husserl, (1962), Chs. 5-6.

2.   Husserl, (1970), esp. "Train of Thought in the Lectures," pp. 1-12 & Lectures II-III, pp. 22-42. Also, for a discussion of the inferred notion of the "object-in-itself" as the Cause of experiences see Husserl, (1962), Section 52, p. 147.

3.   For a full exposition of the Phenomenological Bracketing or *Epoche* see Husserl, (1962), Ch. 3, Sections 30-32, and (1970), esp. Lectures II-III.

4.   See Husserl (1962), pp. 117-20 & 214-5 & Husserl (1970), pp. 8, 53, 56 & 59.

5.   See for example, Husserl (1962), Section 52, p. 147 & pp. 222-3. See also, Husserl (1970), esp. "Train of Thought in the Lectures", pp. 1-12 & Lectures II-III, pp. 22-42.

6.   Kant, B. 139. See also B. 130. My addition in parentheses.

7.   Merleau-Ponty, p. xiv.

8.   Ibid., pp. 36-7.

9.   Note that if all experience is conceptually mediated in the way claimed, then there can be no experience of, nor therefore any empirical evidence for, the existence of such pre-conceptual sense-data etc. However, even if there were, indeed, no such sense-data, "Gestalt switches," (eg. between duck and rabbit, faces and vase, young girl and crone etc.), nevertheless still provide ample empirical evidence of the conceptual mediation of experience.

10.   Quine, p. 44.

11.   Einstein quoted in Heisenberg (1971), pp. 64-5. My addition in parentheses.

12.   Heidegger (1978), pp. 247-8.

13.   Popper, p. 59, Footnote 1. See also p. 107, Footnote 3.

14.   See Babich, p. 149 ff.

15.   Husserl (1970a), p. 96, first parentheses mine, second Husserl's. See also pp. 94, 167, 233. However, while by the time of the *"Krisis"* Husserl has come to recognize this, the insistence of the earlier Husserl on a presuppositionless description indicates that this was not always the case.

16.   Lakatos, p. 130. My addition in parentheses.

17.   Ricoeur (1974), p. 79, my addition in parentheses. See also Ricoeur, (1976), p. 23.

18.   Husserl (1962), p. 214. See also pp. 215 & 331.

19.   Sartre (Undated), p. 41.

20.   For a full discussion of reification and its falsification of the nature of the subject see Ibid., esp. pp. 33-54.

21.   Ibid., p. 53.

22.   Ibid., p. 42.

23.   Indeed while Husserl was arguably equivocal on the notion of a transcendental ego, he himself referred to the subject and object "pole" of experience. See for example, Husserl (1969), p. 66.

24. Margeneau, p. 47.

25. Heidegger (1962), p. 152.

26. Ibid., Section 43, pp. 246-50.

27. Ibid., pp. 95-102.

28. See for example, Foucault, p. 387.

29. Heidegger (1962), pp. 72 & 487.

30. See, for example, Ibid., pp. 83-4 & 235-44.

31. Thus so far from coming to a world initially devoid of value, the world is, in the first instance, full of value and significance. See for eg., Sartre (1947).

32. See Heidegger (1962), pp. 95-102, & 235-44. For a full discussion of this point see, for example, Heidegger (1978a), and Heidegger, (1977), and Husserl, (1970a), pp. 50-51 & 127.

33. Habermas, pp. 316-17. See also p. 171 & 148-9.

34. Ibid., p. 314. See also pp. 148-9 & 171.

35. Nietzsche, 3:526. Nietzsche's addition in parentheses.

36. Note that, as many critics have pointed out, the notion of falsification is structurally or diacritically parasitic upon the notion of truth, and thus Nietzsche may be implicitly affirming what he explicitly seeks to deny.

37. Heisenberg (1958), pp. 8-9.

38. For Husserl on the relativism of the life-world see (1970a), pp. 138-9.

39. Sartre (1956), p. 308.

40. Neils Bohr quoted in Heisenberg (1971), p. 88.

41. See Husserl (1970a), pp. 138-9.

42. See Note 4 above for example. For a full discussion of the phenomenological notion of essences, see Husserl (1970), pp. 52-56.

43. Russell, p. 163.

44. See quote 12.

45. Heisenberg (1971), p. 11.

46. Capra, pp. 71-2.

47. Sartre (1956), p. 307.

48. Heidegger (1977), p. 173.

49. Charles Wheeler as quoted in Mehra (1973), p. 244.

50. de Saussure claims that syntax generates semantics. See de Saussure.

51. Bohm, pp. 161-2.

52. Heisenberg (1971), p. 85.

53. Heidegger (1962), p. 414.

54. Husserl (1970a), p. 51.

55. Stapp, H.P. quoted in Capra, p. 143.

56. Capra, p. 228.

57. Weyl, p. 171.

58. Capra, p. 313.

59. Chew, G., Gell-Mann, M., & Rosenfeld, A. quoted at Ibid.

60. Bohm, D. & Hiley, B. quoted by Capra, pp. 141-2.

61. Habermas, pp. 78-9.

6 Observing the Analogies of Nature

Daniel Rothbart

Introduction

It is a truism within science that modern instruments are central to most experimental studies. The "success" of modern science in general must be credited to the capacity of instruments to provide human contact to the inherent structures of physical reality. Is such optimism philosophically naïve, in light of the widespread critiques of foundationalist philosophies of science? In my opinion it is not, but only if we overcome the empiricist biases of instruments functioning as sensory magnification.

Traditional empiricist epistemologies have raised major philosophical obstacles to a clear understanding of instruments. Empiricist philosophers have given scant attention to instruments as a separate topic of inquiry because the use of instruments is presumably epistemologically reducible to the epistemology of common sense experience. Instruments function to magnify our physiologically limited sensory capacities by "causally" linking the specimen's sensory properties to accessible empirical data; such data in turn are validated by the same empiricist standards used to access ordinary (middle-sized) phenomena. Thus, no epistemic insight is revealed by studying instrumentation *per se*, according to empiricists.

From both sides of the Atlantic Ocean, critics of empiricism have championed the cause that all sensory data are theory-laden. Yet even this dictum has the effect of minimizing the philosophical significance of instruments, primarily because many philosophers are working within the empiricist categories which distinguish the subjectivity of theory from the apparent objectivity of data. Such a distinction assumes a scientifically naïve design of scientific instruments. Once we overcome simplistic conception of instrument design, the theory-laden character of data will not imply the inherent failure (subjectivity, circularity, or rationalization) of instruments to expose nature's secrets. Rather than a warrant for its subjectivity, this theory-laden character of data reveals the instrument's proficiency.

I argue that the success of instruments is based on their design as analogical replication of natural systems. The designers of modern instruments create artificial technological replicas of familiar physical systems. Progress in designing proficient instruments is generated by analogical projections of theoretical insights from known physical systems to unknown terrain. Instrumentation enables scientists to expand their limited theoretical understanding to previously hidden domains.

After exploring this analogical function of instruments (Section 1), the nature of instrumental data is discussed (Section 2), followed by an explicit rejection of both skepticism and naïve realism (Section 3). In the end I argue for an experimental realism which lacks any theory-neutral access to the fundamental analogies of nature.

Instruments as replicas of nature

What is the function of scientific instruments? A rather sophisticated reformulation of empiricist doctrines, within the philosophical framework of evolutionary epistemology, appears in Robert Ackermann's *Data, Instruments, and Theory*.[1] Ackermann argues that the primary function of scientific experiments in which instruments are used is to break the line of influence from theory to fact, that is, to constrain the subjectivity of interpretation by the intersubjectivity of fact.[2] Instruments depersonalize experimentation; no single scientist can legitimately alter the experimental results.

But Ackermann's rationale oversimplifies the character of instrumental design, and the function of instruments. Contrary to the empiricists' claims, extraordinary phenomena become ordinary not by sensory magnification but rather by projection of powerful theoretical insights to previously unexplored terrain. A modern scientific instrument is designed as a complex system of many action/reaction events; the specimen is "activated" by physically responding to humanly generated prodding, pushing, and poking within the experimental setting. Rather than observing the specimen's properties in its passive state, the experimental scientist detects the physical performance of the specimen under experimental stimuli. Such a performance is called the experimental phenomenon of interest.

The experimental phenomenon is a local event of interaction, that is, a moment of intersection between the specimen's structure and humanly designed conditions. (Even the term "between" retains the Cartesian metaphor of distinctly individuated domains.) Such an event is neither generated purely by external physical structure nor purely by internal conceptualization. Thus, in a certain sense, all experimental properties are tendencies, or conditional manifestations of the specimen, to react to certain experimental stimuli. The specimen has tendencies which are manifested only if certain humanly designed experimental conditions are realized, as Rom Harré writes.[3]   Although such conditions are teleologically determined, the tendencies are grounded on the specimen's real physical structure, independent of social convention. Both naïve realism and radical skepticism are grossly inadequate to capture the success of modern instruments.

Upon what reasonable basis can scientists trust modern instruments to provide such dispositions of the specimen's structure? The reliability of modern scientific instruments rests on the analogical underpinnings of instrumental design. Within instrumental engineering, such a sequence of real physical processes is intended to duplicate select patterns found in natural phenomena. The instrument's designers typically dissect, restructure, and reorganize natural systems for the purpose of replicating select symmetries of nature. The physical sequence of events from specimen structure to data readout constitutes a technological analogue to multiple natural systems, based on underlying causal models of real world phenomena. Such models are necessary to generate, or at least aspire towards, a reliable information-carrying signal from the specimen to the data. Thus, the instrument's success at exposing unknown properties of nature is tied directly to the capacity of scientists to extend analogue models of natural events to certain artificial contexts. In this context the instrument can expose previously hidden physical properties by analogically extending known physical symmetries to the specimen under investigation.

One major task for any designer is the selection of the most promising analogical system to function as the generator for instrument's relevant causal relations. The theoretical conception of the instrument's design and operation rests on such analogue systems. The analogical origins of such designs become hidden under the cloak of repeated experimental successes, but the importance of such analogical systems becomes most conspicuous at the time of the instrument's design. Consider two examples.

1. C.T.R. Wilson designed the cloud chamber not as a particle detector but as a meteorological reproduction of real atmospheric condensation. Meteorology in the 1890s was experiencing a "mimetic" transformation, in which the morphological scientists began to use the laboratory to reproduce natural occurrences.[4]   The mimeticists, to use a term by Peter Galison and Alexis Assmus, produced miniature versions of cyclones, glaciers, and windstorms. John Aitken's dust chamber, recreating the effects of fogs threatening England's

industrial cities, directed Wilson's design of the cloud chamber.[5]   Wilson transported the basic components of the dust chamber (the pump, reservoir, filter, values, and expansion mechanics) to his cloud chamber for the reproduction of thunderstorms, coronae, and atmospheric electricity.

J.J. Thompson and the researchers at the Cavendish laboratories gave the "same" instruments not only an alternative function as particle detector but also a new theoretical rationale. Rather than imitating nature, Thompson intended to take it apart in exploring the fundamental character of matter.[6] For their matter-theoretic purposes, scientists at Cavendish became indebted to Wilson's artificial clouds for revealing the fundamental electrical nature of matter. Galison and Assmus write "As the knotty clouds blended into the tracks of alpha particles and the 'thread-like' clouds became beta-particle trajectories, the old sense and meaning of the chamber changed."[7] For 20th century physicists the formation of droplets were replaced by the energies of gamma rays, the scattering of alpha particles, and discovery of new particles. Wilson and the matter physicists proffered rival theoretical interpretations of the chamber's causal structure, interpretations derived from distinct physical analogues. Thompson and Wilson employed different instruments from one another.

2. The modern absorption spectrometer is commonly used for identification, structure elucidation, and quantification of chemical substances. A beam of electromagnetic radiation emitted in the spectral region of interest passes through a series of optical components such as lenses and mirrors. The radiation then impinges on a sample. Depending on the nature of the sample, various wavelengths of radiation are absorbed, reflected, or transmitted. That part of the radiation which passes through the sample functions as the sample's fingerprint. The radiation is detected and converted to an electrical signal. The electric output is electronically manipulated as desired and then sent to the readout device, which may be a meter, a computer, a controlled video display, or a printer/plotter.[8]

As the primary focus of attention for experiments, the signal is defined roughly as an information-carrying variable. But what is the physical composition of the signal? The signal is understood as discrete photons, by analogy to the electromagnetic impulses from a flash of light. When a flash of light is observed with a photomultiplier and displayed on an oscilloscope, the observed signal becomes a set of impulses. For an absorption spectrometer the photoelectric signals are physically analogous to the radiation impulses of electromagnetic radiation based on models of electromagnetic radiation.

114

Data structures

How can we explain the reliability of scientific data in light of the post-modernist and anglo-american revelations concerning the theory-laden character of instrumental output?

From a praxis orientation the reliability of instrumental data rests on a wide ranging set of practices within the instrument's design, function, and operation. Ian Hacking's dictum that we see with a microscope and not through one[9] applies perfectly to spectral analysis. For many modern instruments the benchmark for the data's reliability is almost never direct visual perception, except for the obvious purpose of manipulating the apparatus. Within spectral analysis the prevalence of visual data, e.g., the yellow from a sodium flame, has been replaced in modern spectrometers by discursive readouts. Typically, the experimenter reads the graphic displays, the digital messages, or the coded charts directly from the instrumental readout. The computer controlled video display and the more common printer/plotter, for example, always convey information through a language that is accessible to the trained technician. For example, a photomultiplier readout device transforms the radiant energy of a signal into electrical energy, while simultaneously multiplying the generated current approximately one million times. It would even be possible for data to emanate from a computerized synthesizer producing an audio readout in English sentences, although such readouts would likely be inefficient.

Trivially, such digital readouts for modern instruments include a sensory component, but the sensory component in any act of reading information is not the epistemic grounding of the words' meaning. The access to data from certain instruments, such as the bubble chamber, do require the perceptual act of "seeing" the image on photographic plates. But frequently the tracks produced take place much too quickly to be detected by an unaided visual system, and many trajectories of electrically neutral particles produce no visual record at all. Moreover, the data analysis frequently involves complex measurement techniques, with various statistical and curve fitting procedures. Visual images are far less relevant for data reliability than are lengthy and complex procedures of information extraction.[10] Such instruments constitute readable technologies because they communicate information directly to skilled "observers," as Patrick Heelan argues.[11]

Returning to the example of spectrometers, data analysis typically includes reading series of spectral lines. But the very determination of the number of spectral lines requires active manipulation of the apparatus, rather than the passive mental inspection of the context of sensory experience. The sharpness of the image is determined by the instrument's resolving power, which represents a limit of the separation of wavelengths into distinct images. How much drop in intensity must there be between adjacent spectral peaks for the lines to be considered resolved? Lord Rayleigh required of resolution an intensity drop of at least 19% between peaks. But any of several answers may be given, although most scientists require at least a 10% drop between peaks. In practice scientists operationalize the notion of resolution by actively adjusting the calibrations, manipulating the instrument's components, and changing the meter readings.[12]

No instrumental data are secured to a non-patterned bedrock of empirical attributes. The information conveyed by data for most instruments is inherently structured into units of relatively low levels of abstraction. In general, facts are created by applying fine-grained conceptual systems, *ceteris paribus*, to the coarse-grained deliverance of our perceptual systems.[13] Such information is conveyed by data structures, which are localized conceptual representations of the detected signal. The discovery of a data structure reflects the twists and turns of evolving scientific thought. Like any causal model of physical processes, each data structure is the result of, and a record of, the past. Simply put, a data structure extends the range of applicability of external theories to unexplored terrain. Trivially, data structures are theory-laden, but this declaration should not be used to resurrect the empiricist's issues surrounding the purity, or lack of purity, of sensory perception.

The construction of a particular data structure is conceptually and causally driven by analogical projections from external data structures, typically found in nature. The conceptual grid defining each data structure supplied by analogical connections to distinct structures. The fine tuned conceptual grid of each data structure is partially generated by analogical projections from some distinct instrumental findings, as Harré has shown.[14] Consequently, such analogue structures are data-constituting when the analogues define the limits and possibilities of an instrument's informational capacity.

Data-constituting analogies are quite prevalent throughout the history of optics. In his 1675 "Hypothesis of Light," Newton describes experiments on the "musical spectrum of light." As Penelope Gouk documents,[15] he discovers that the light spectrum can be divided into the same arithmetic ratios underlying a particular musical scale. Although initially categorizing light into five major colors, the musical scale of light motivates Newton to add two more colors. The spectrum's division into seven colors became obvious to Newton from the analogy to the seven tone musical scale, expressing the traditional names for a rising major scale from an eleventh century octave. Thus, both light and sound are caused by motions of matter transmitted through a medium. Just as the musical sounds are caused by periodic vibrations of varying wavelengths in the air, so too the sensations of colors are produced by aethereal vibrations. The intensity of vibrations generating red is causally analogous to a particular tone, "for the Analogy of Nature is to be observed."[16]

116

It is the moment of such analogical projections that generates possible data structures; we locate the formation of an instrument's possible data structures at the point of discovering powerful conceptual grids from distinct analogue systems.

Overcoming the skeptic's noise

This analogical conception of instrumentation does not warrant a skepticism about the capacity of instruments to reveal the specimen's physical dispositions. The skeptic is correct, although trivially, that no signal is completely unequivocal, that no mapping from data structure to specimen structure is one-to-one, and that the signal-to-noise ratio is never indefinitely high. Nevertheless, reliable channels of communication, based on background theories, can be in principle achieved so that the signal is practically unequivocal, that the mapping from data structure to specimen structure approaches one-to-one, and that the signal-to-noise ratio can be maximized. The images from infrared detectors employed by astrophysicists to reveal newborn stars are not the complete fabrication of the experimenter's symbol system. The triple line sequence of spectral analysis is not an artifact of the scientists' conceptualizations. The tracks of alpha particles within a bubble chamber are not fictitious concoctions by self-deluding scientists. A skeptic shows signs of neurosis if the channel conditions underlying the instrument are repeatedly checked beyond necessity.[17]

Furthermore, the analogue system functions as an idealized prototype that is projectible onto the phenomena under scrutiny. Technological obstacles to instrumental progress frequently require radical conceptual breakthroughs. Such breakthroughs transform the "inconceivable" into the manifestly plausible by overthrowing antiquated scales of conceptual vision, that is, by employing powerful new analogies. Such analogue systems acquire normative force for understanding the entire domain of inquiry. Yet the discovery of fresh analogies, and new prototypes, does not always require a monolithic overhaul of the entire scientific enterprise, as is suggested by a Kuhnian-type paradigm shift. Newly discovered analogies typically yield a specifiable and localized transformation of some problematic subject. A radical discontinuous transformation of physical science is not a necessary result of such analogical transformations.

Corresponding to such transformations are profound alterations in the norms of scientific language. Bold theoretical insights frequently create a crisis of scientific vocabulary that can be circumvented only by non-literal expressions. The metaphoric origins of many scientific descriptions become most evident at the moment of theoretical insight, such as the clock-related properties of human behavior, the rate of flow attributes of electrical current, the information

117

processing characteristics of the mind, the wave-like properties of light.

Simultaneously, a prominent factor in judging a theory's success is its capacity to motivate instrumental progress. The theory's fertility is partially measured through the theory's capacity to generate instrument designs. Thus, mutual dependence arises between instrumental design and theoretical progress: the instrument's design requires the complex combinations of various theoretical insights, and the theory's vitality is measured by the design of successful instruments. In this respect the internal/external distinction assumed above between the specimen's unknown parameters and the background theoretical models must be qualified.

Conclusion

The radical skeptic cannot explain the success of scientific instruments. For both infrared detector telescopes exploring newly discovered galaxies and superconducting super colliders detecting bosons and neutrinos, the instrument's signals span extraordinary epistemic distances to a point within the experimenter's immediate reach.

Nevertheless, instrumental success inevitably reflects the human intervention of technological manipulation, conceptual segmentation, and implicitly historical thought, for the immediate purpose of producing an information-rich physical reaction. This human intervention is not simply a practical necessity for using instruments, but defines the very subject matter of scientific inquiry in general. Again, scientific phenomena are events of interaction from which the specimen's properties are dispositions released under specifiable environment conditions. Within instrumentation such conditions are artificially generated. No pristine contact to the undifferentiated flow of physical reality is possible. The very domain of scientific inquiry is a sphere of physical action which is decidedly anthropocentric, reflecting the teleological needs and capacities of human agency.

Consequently, the philosophical foundation of convergent realism is undermined. The convergent realist promotes the scientific ideal of validating a unified ontological system, in order to capture the structure of an undifferentiated physical reality. Such an ideal has been subject to the charge that no theory-neutral methodological measurement of rival theories is within reach. Moreover, such a pure undifferentiated world is not even the subject matter of science. Again, scientific theorizing strives towards a systematic replication of the physical tendencies which are released under humanly generated experimental conditions. Such conditions are themselves dispositional properties which are exposed from some previous study. Human agency becomes a pivotal dimension of the very domain of science at every level of inquiry. The interactionism within instruments is paradigmatic of scientific research generally. Throughout the history of science the conceptual foundations and premises of theories undergo continuous and sometimes revolutionary transformation, as Mary Hesse shows.[18]

The subject matter of modern experiment is neither purely constructed nor passively discovered. Nature's secrets are exposed by the continual juxtaposition of distinct theoretical systems, a juxtaposition that promotes a perspectivism driven by powerful analogies of nature.

Notes

1.   Ackermann, R. (1985), *Data, Instruments, and Theory*, Princeton University Press, Princeton.

2.   Ibid., p. 128.

3.   Harré, R. (1986), *Varieties of Realism*, Basil Blackwell, Oxford, Chapter 15.

4.   Galison P. and Assmus, A. (1989), "Artificial Clouds and Real Particles," in *The Uses of Experiment*, Gooding, D., Pinch, T., and Schaffer, S. (eds.), Cambridge University Press, Cambridge, pp. 225-274.

5.   Ibid., p. 265.

6.   Ibid.

7.   Ibid., p. 268.

8.   I thank Professor Suzanne Slayden for providing the background information on spectrometers.

9.   Hacking, I. (1983), *Intervening and Representing*, Cambridge University Press, Cambridge, Chapter 11.

10.   Woodward, J. "Data and Phenomena," *Synthese*, 79 (1989): fn. 35, 459.

11.   Heelan, P. A. "Natural Science as a Hermeneutic of Instrumentation," *Philosophy of Science*, Vol. 50, (June 1983): 181-204.

12.   "Optical Qualities of Spectroscopic Instruments," in *Encyclopedia of Spectroscopy*, edited by E. G. Clear, pp. 244-245.

13.   Harré, R. *Varieties of Realism*, op. cit., p. 176.

14.   Ibid., p. 175.

15.   Gouk, P. "Newton and Music: From the Microcosm to the Macrocosm," *International Studies in the Philosophy of Science: The Dubrovnik Papers I*, Vol. 1, No. 1 (September 1986): 36-59.

16.   Newton, I., "Newton to Oldenburg: 7 December 1675", in H. W. Turnbull, H. W. (ed.), (1959), *The Correspondence of Isaac Newton, Volume I: 1661-1675*, Cambridge University Press, Cambridge, p. 376.

17.   Dretske, F. (1981), *Knowledge and the Flow of Information*, MIT Press, Cambridge, pp. 115-116.

18.   Hesse, M. (1980), *Revolutions and Reconstructions in the Philosophy of Science*, Indiana University Press, Bloomington, p. 174.

7 Natural Science is Human Science. Human Science is Natural Science: Never the Twain Shall Meet

Charles Harvey

Introduction

The title of my paper plays upon an equivocation inherent to our current categorization of the sciences. The equivocation leads to paradox; the paradox demands decision. For to say that "natural science is human science" is in one sense to utter a trivial truth; for obviously, it is the human being who does natural science; but in another sense, it is to utter what appears a conundrum, for we often mean by "natural science" precisely the antithesis of "human science." A similar trivial truth and a similar conundrum arise if we reverse the claim and say that "human science is natural science." In the trivial sense, human science is surely "natural" for it has proven "natural" for the human being to do human science, while in the perplexing sense, we often mean by "human science" precisely the antithesis of "natural science."

In this paper I work through the senses of "science" embedded in each of these statements. In section 1, I show how each statement has both an act-noetic and an object-noematic sense, resulting in four logically possible meanings for the two claims. In section 2, on the basis of the sense-analyses of these phrases, and through the use of imaginative variants, I describe how each type of science might well make claims to priority and subsumption of the other type. In section 3, I argue that on their respective ways to totalizing reduction, each type of science passes by the other, never touching, never merging, never making cognitive contact. Finally, in section 4, I argue that while philosophers should record the moment of passing – the point at which the natural sciences pass through the sense-barriers marking the domain of human scientific subjects and the point at which the human sciences pass through the sense-barriers marking the domain of natural scientific objects – they should resist legislating against the scientific transgression of normalized sense-borders. In a postmodern world, I argue, we are best off letting the sciences methodologically totalize, while teaching ourselves a phenomenological calm about life-worldly sense-gambits, and an existential agility about what we count real and when we do so.

Natural and human science

"Natural science is human science": act-noetic interpretation, sense$_1$

Natural Science is a science done by human beings – in this sense natural science is a human science. This is the act-noetic sense of the claim that natural science is a human science because this assertion means that the cognitive and manipulative achievements of natural science just are the cognitive and manipulative achievements of the human being. Natural science is a human science because *we* do it; natural science is our achievement. To say that "natural science is a human science" is to say that doing natural science is an existential characteristic of human being. This is the sense in which the claim "natural science is a human science" is a trivial truth. But this truth need not remain so trivial, it can quickly become radically reductive.

"Natural science is human science": object-noematic interpretation, counter-sense$_1$

If the claim that "natural science is human science" strikes one as a trivial truth when interpreted from the act-noetic perspective, it strikes one as a conundrum, as noematically explosive, when given its sense from the object-noematic perspective. The object-noematic sides of natural science and human science refer, on the basis of pre-determined acts of sense-circumscription, to radically distinct object-domains. For instance, a science is a natural science precisely because its subject matter is always an "object." The subject matter of natural science is precisely the non-human, non-personal, non-subjective; indeed, in the fulfillment of its ideal of success, the subject matter of natural science becomes precisely a "non-subject," it becomes an object *qua* object depleted of all and any trace of the subject.[1] The object of natural science must be stripped of all psychic, conscious, distinctively human characteristics. In reference to the object-noematic interpretation of the claim that natural science is a human science, then, we have a counter-sense, an unhappy noema. Hence, from the object-noematic perspective, the claim that natural science is a human science is a counter-sense.

Human Science is a science done naturally by human beings: in this sense human science is natural science. This is the act-noetic sense of the claim that "human science is a natural science" because this assertion means that as natural beings living together in community it is natural for human beings to study human beings, to do human science. Insofar as a consciously motivated change is bought about in any social-psychological situation, humans have engaged in human science. In its basic forms, human science is the most natural of sciences: it stems from a natural situation that promotes rational reflection and action, and these in turn (naturally) produce change. To act on the basis of reflection, so as to promote (planned) change in a social-psychological situation, is to have engaged in human science. And for human beings, such reflection, action and promotion of change is a natural activity. It is in this sense that the claim "human science is a natural science" is a trivial truth. But this truth need not remain so trivial, it can quickly become radically reductive.

"Human science is natural science": object-noematic interpretation, counter-sense₁

If the claim that human science is a natural science strikes one as a trivial truth when interpreted from the act-noetic perspective, it strikes one as a conundrum, as noematically explosive, when we attempt to render it meaningful from the object-noematic perspective. Again, the object-noematic sides of human science and natural science refer, on the basis of the pre-established sense-circumscription of their regions, to radically different types of objects. From this sense-circumscribed perspective, a science is a human science because its objects are inherently meaningful, inherently laden with human significance, inherently a product of human activity. The subject matter of human science is precisely the "subjective" insofar as the subjective inheres in the body, in cultural objects, in cognitive, emotive and physical achievements. Hence, to make the object (read: subject) of human science a mere object is a counter-sense. To strip the human element from the objects of human science is to forsake the human in human science. From the object-noematic perspective, then, the claim that human science is a natural science is a counter-sense, an unhappy noema.

Totalizing tendencies

Motivating the senses of "science" thus delineated are the priority and reducibility beliefs held by each of the scientific perspectives considered. It is deep allegiance to the totalizing possibilities of a science that generate the counter-senses described, just as it is deep allegiance to the intrinsic sense of the subject matters of each science that generates resistance to those counter-senses. In this section I suggest how each type of science might indeed subsume the other as a branch of itself. Interestingly, this could be done simply by extending the insight embedded in each of the trivial truth claims of section 1 to the point that these claims become the counter-senses of that same section. I will try to demonstrate how, historically, this might have occurred.

The epistemic and ontological priority of human science

The claim that the human sciences have epistemic priority over the natural sciences begins by reference to the mode of access necessary for the achievements of any science. It is argued that because the starting point for the achievements of any science is the cognitive situation of human being, human being must be understood before we can understand the human understanding of the natural world.[2] And it is the role of the human sciences to articulate both the human understanding of its first order situation and that first order situation itself.

This epistemic priority of the human sciences is correlated with the ontological priority of the human world to any other "world." If the human sciences have act-noetic (or epistemic) priority over the purely natural sciences, it is because the object of the human sciences, the human world, has object-noematic (ontological) priority over all other "worlds." Here it is claimed that for any scientific theory, the human world is the "premise-world."[3] All higher-order worlds, all theories, stem from it and must, ultimately, refer back to it. If a theory does not refer back to the lived world, it has "lost touch" with reality; to use philosopher's language, "bridge-principles" for the theory are lacking, and hence, so too is the value or usefulness or reliability of the theory.

Claiming that these priority perspectives are necessary for the possibility of knowledge is a form of the "familiarity thesis."[4] The familiarity thesis claims that all new knowledge is knowledge discovered, constructed, in terms of old knowledge, in terms of things already known or understood. Heidegger adopts this position in *Being and Time* with his thesis of fore-structured understanding,[5] as does Dewey in *Experience and Nature*.[6] The physicist N. R. Campbell, with his claim that in order to make sense, scientific reduction must always "display an analogy" to what is already known and familiar is also advocating the familiarity thesis.[7] When this thesis is pushed to its extreme, it necessarily privileges the everyday world and its conceptual frames, and thereby lays the ground for the reducibility of natural science to human science.

It is common parlance to say that whereas the natural scientists seek to explain, the human scientists seek to understand. This distinction between understanding and explanation, however, is itself predicated upon the deeper distinction between intentionality and causality. If the natural sciences rely upon physicalistic causality as the human sciences rely upon intentionalistic motivation, and intentionalistic motivation is shown to be prior to causal relationality, then natural science will be shown to be posterior to, because ultimately explainable in terms of, human scientific motifs. The basic explanatory terms of the human sciences, intentionality and motivation, will be necessary to explain the basic term of the natural sciences, "causality."

Husserl, of course, claims that motivation is the most fundamental form of intentionality.[8] It refers to the implicatory structure of all experience. As early as the *Logical Investigations,* he noted that it is the motivational structure of experience that prompts one, even without reflection, to 'infer' things about the "not-given" and the "will-be-given" from the given.[9] But if this is also what causal understanding allows us to do, then causal relations must ultimately be understood as motivational ones.

Such an argument was given its most notorious, though inchoate, form by David Hume when he reduced causal relations to inductive relations and inductive relations to associative ones.[10] A move not too different from Hume's lies at the heart of the reducibility thesis of the phenomenologically interpreted human sciences. Here it is claimed that when human experience is understood genetically, pure receptivity functions as the lowest layer of experience. This receptivity is not passive, however, for the passive data that affect the subject are one with the triggering of the associative-motivating processes of human consciousness.[11] Once the associative-motivating processes of consciousness "kick-in," the conditions for the possibility of induction are established.[12] The experience of 'x,' then 'y', quickly becomes the expectation of 'y' whenever 'x'. *Constantly confirmed* expectation is soon objectified, naturalized, and comes to be called "causality." Because consciousness tends to objectify any of its achievements, the noetically rooted "I expect 'y' to follow 'x'" is translated into a noematically structured proposition "If 'x', then 'y'." This proposition is cast into the world and from then on is a constitutive factor in, and a constituted layer of, the world.

All the problems concerning the justification of induction, and the use of induction as a basis for causality, thus arise. Husserl, however, similar to Dewey, realized that "induction" is as intrinsic to our experience of the world as is sunset and sunrise – more so, in fact, because the latter is meaningfully constituted only by means of the former. Indeed, Husserl realized that the Kantian category of causality was itself a constituted *ideal objectivity* on the noematic side of experience that arose from the repeated validity confirmations of associative, motivating, inductive experiences.[13] Moreover, because the category of "causality" is a constituted, ideal noematic pole for the organization of experience, any particular causal claim must have first passed through the gauntlet of associative-motivating processes and repeated inductive confirmation before it was translated from the subjective expression "I expect 'y' whenever 'x'," into the objective proposition, "When 'x', then 'y'." In these manners, then, "causality" can be viewed from the eyes of the human scientist as an intentionally constituted validity achievement of the conscious subject.

By way of a thought experiment that might help support the reducibility thesis of the human sciences, consider the following alternative history for the development of our current sciences. Imagine that the dream of many modern philosophers had played true and that our cognitive achievements had progressed *from* an understanding of the knowing subject *to* an understanding of the objects known by that subject, rather than the other way round. Rather than first gaining some knowledge about distant reaches of the universe, then about our physical planet, then about other animals, and finally about ourselves, imagine that we had first gained some scientific knowledge about ourselves and our modes of knowing, then, progressively, about all those things at a distance from us. This would have fixed and fulfilled the early epistemologist's broken dreams by way of fulfilling the belief that we must know what and how the human subject knows before we can know if what it knows is really so. How might such an alternative historical progression of knowledge have changed our dominant epistemic paradigms?

If the progression of knowledge had occurred in this fashion, we might understand our universe in terms of motivational processes rather than causal ones. We might understand the relations of purely physical bodies in our universe as motivational relations, perhaps even intentional ones. In this vein, perhaps, we would have never built the barrier between mind and world that we still seek to overcome. The laws of the "world" would be the laws of the "mind," and vice versa. Indeed, the radical distinction between the two may not have occurred.

Such an occurrence might have been a very real possibility. Scholars of ancient thought suggest that Anaximander's concept of "law," (in reference to the "laws of the cosmos" or "nature"), was an extension of the laws of the city-state to the universe at large.[14] That is, the first "laws" of nature really were laws – laws of the type humans create; intentionally and motivationally generated laws. Similar to these first proto-scientific laws of Anaximander, it might be argued, all of nature's laws have arisen from us. The achievements of natural scientists are the achievements of human scientists: the "laws" of nature are grounded in the laws of *our* nature. The reducibility thesis of the human scientist, then, would take these claims as its fundamental postulate and regulative ideal, and would have as its task, proving that all natural scientific laws could be reduced to human ones, all so-called natural scientific objects to human scientific constructions, to socially generated constituting acts.[15]

If this reducibility thesis of the human sciences were granted, if we had moved slowly and systematically from the trivial truths of section 1(i) to a universal science of all things, then would the counter-sense of section 1(ii) be salvaged, made into sense? Under such conditions, interpreted from an object-noematic perspective, would natural science be a human science? Was it merely the contingencies of history that kept this from happening? Or, rather, is there something real about ourselves and the world that demands our great regions – the regions of mind and matter – be what they are? I will address this question in the third section.

Let us turn now to the priority and reducibility thesis of natural science over human science.

The epistemic and ontological priority of natural science

The thesis that the natural sciences have epistemic priority over the human sciences begins by reference to the apparent fact that all mental and spiritual realities are grounded in physical, material realities. In the absence of the latter, the prior are impossible. If this is so, then all motivational and intentional relations are grounded in, perhaps reducible to, causal relations. Hence, the ontological priority of the natural sciences to any other type of science is correlated with the dependence of all forms of reality on material reality.

In the case of the claim that natural science is prior to human science, causality, the basis of motivation, my examples need not be extensive, nor as imaginative, as in the case of the reverse claim. A free imaginative variant on history is hardly needed because much of our Western, scientific-philosophic tradition has been one long attempt to demonstrate just this thesis. Indeed, it is a thesis that has for the most part been assumed to be true.

It emerged first with Leucippius and Democritus; the epistemic garb in which it has cloaked its regulative intentions has changed little since then. Leucippius, Democritus, and then at the threshold of modern science, Galileo, Hobbes, Descartes, Locke and many others have claimed that qualitative differences between things are reducible to, explainable in terms of, quantitative, purely material differences. "Shape," "arrangement" and "position" of the "atoms," Leucippius and Democritus are claimed to have claimed, can account for all the qualitatively perceived differences we have of the world. "Color does not exist in nature;" Democritus said, "for the elements – both the solids and the void – are without qualities."[16] So it is, too, with the other senses; so it is with the human intellect, passions, and socio-cultural creations. All can be understood as "epiphenomena," as causal consequences of the material workings of Nature. "Man is a machine," said La Mettrie,[17] and from the workings of the machine comes the ever apparent but ever unreal rainbow-stage upon which the drama of human life is played out. The social-psychological-cultural colorings that make up the life of humankind, can be, and one day will be, explained in terms of its causal, material sources. We have discovered that water is really H_2O, that diamonds are coal; so too shall we discover, to paraphrase Marx, that intentional life is secreted from brain life as is bile from the liver.

This story is so well known to philosophers that I won't rehearse it in detail.[18] I will, however, suggest one reading, one imaginative variant, on how the human scientific, intentionalistic reduction developed in the previous subsection could be turned around to lend support to the causal reduction of the natural scientific attitude. From there, the causal reading of the world could be taken all the way up to the most intangible spiritual realms of social-psychological life.

Again, we take Hume as our springboard. The "gentle force" of association may be naught but the unconscious force of stimulus-response made consciously manifest. And indeed, that is just what it was for Ivan Pavlov,[19] and what it became for John B. Watson[20] and B. F. Skinner.[21] Stimulus-response is a body-brain habit, a mechanistic event; upon it rides, says the materialist, the associative-motivational nexus of human psychic and spiritual life. From the reflexes of the body-brain come the mind's "gentle forces," its associative and motivational "reflexes," and finally its "intentions." Rather than reasoning as we did earlier, we would now argue this way: Because of cause and effect relations that exist in the material world independent of consciousness, when 'x' causes 'y' again and again, the experience of 'x' comes to *cause* the expectation of 'y' in mental life. The associative-motivational processes "kick-in" just because they were caused to do so by worldly causal relations that caused brainly causal relations that caused associative (really, causal) relations. From *the fact that* 'x', then 'y' again and again, when we experience 'x' we expect 'y' because 'x' in our brain produces 'y' in our brain. From the material, causal reality, 'x' then 'y', the world has cast into our brains and hence into our "minds" the expectation that "If 'x', then 'y'." But the human subject, thereby, has been made a constituted layer of the causal, material world.

While neither we nor anyone else can yet start with simple causality and "go all the way up" to humankind's highest spiritual realms, the causal, material source has now been placed right inside our most basic mental activity – the fundamental, most pervasive mental activity shared by babbling babes and brainy behaviorists alike – *association*. It is only a matter of time, the totalizing natural scientist might say, before not only our hopes and our dreams, our angst and aversions, will be materially explained, but so too will be religion and art, philosophy and love. We have already heard calls to such interpretative programs.[22]

If this reducibility of the natural sciences were granted, if we continue to move slowly and systematically from the trivial truth of section 1(iii) to a universal science of all things, then would the counter-sense mentioned in section 1(iv) be salvaged, made into sense? Under such conditions, interpreted from an object-noematic perspective, would human science become a natural science, because the object (the "subject") of human science would be immediately known as a causal consequence of the natural world? Has it been due only to our slowly vanishing ignorance that this condition has not yet been realized? After a time, could such a sense of ourselves become automatic and natural? In the next section of this essay I will reflect on this question, as well as the one we posed at the conclusion of part 2: Has it been merely the contingencies of history that, so far, have kept this from happening? Or is there something real about minds and things that demands our great ontological regions be what they are?

Sciences that pass in the night

It is at least imaginatively and methodologically possible that all causal relations be understood as intentional ones; all intentional relations as causal ones. It is conceivable that the human sciences be reduced to the natural sciences or the natural sciences to the human ones. Either, sticking fast to its methods, could proclaim explanatory totalization. Yet in spite of great hopes and numerous calls to arms from each camp, it just hasn't happened. Why not?

In its totalizing tendencies, its push towards total reduction, each type of science founders at precisely the junction where ordinary life delimits the ontological regions separating mind-life from pure material being. This is manifest in the feeling of counter-sense, of "exploding" noemata described in part 1. The cognitive dissonance that we feel when reduction goes "too far" reveals the regional demarcations within which we currently think types of science appropriately operate, i.e., the limits within which they can reveal, and not distort, the essential traits of their identified subject matters. At the moment that the trivial truths about minds or matter becomes a counter-sense, we have broken through the sense-boundaries of our regional ontologies. One can argue, of course, that science always has and always should break through such boundaries, that science done honestly, i.e., radically, must break through some of these. Perhaps so; I will, to some extent, support this claim shortly. However, as long as reducing objects from one region to another disrupts noemata, savages *Sinne*, we cannot make good sense of these reductions. Indeed, citizenship in the great regions of being accepted in common life, and demarcated by Husserl in

129

Ideas II, may mark essential rigidity restrictions, the irreducible aspects of beings, through which we cannot reduce and simultaneously maintain sense.

This problem might be articulated in terms of another dichotomy that, however, rather than helping to resolve the questions I've posed, reinforces the distinction that spawns the problem. The dichotomy is between "function" and "essence." Functionally, each type of science can reduce right through the defining ontological borders established and legitimated in daily life. To take an example, in psychology, the cusp discipline *par excellent*, pharmaceutical, i.e., physicalistic, treatment can indeed change intentional life – make one happy or sad, sensitive or numb, vital or lethargic. Similarly, intentionalistic treatment (commonly called "cognitive therapy") – coming to know that and why one is who and how one is – can change one's physical, material being; it can raise or lower blood pressure, change our sleep patterns and sexual appetites, and even change our body's sense of hunger or satiation. And yet, when such changes are made, we continue to *conceive* of ourselves in terms of the ontological categories with which we started – minds and bodies, intentions and causes, spirit and matter. The regional realities clearly affect one another, yet never do they semantically translate or combine, never do they "*Sinnfully*" merge one into the other.

Said another way, even when reduction is functionally successful at penetrating an ontological border and thereby changes and modifies something in the other ontological region, we continue to conceive of the region as separate – ontologically distinct. The totalizing and reductionistic success of a type of science at a functional, practical level, then, has little or no impact on our comprehension or ascription of defining traits of the ontological regions. Like ships that pass in the night, so, too, do these types of science. Each type of science may indeed reduce right through the sense-circumscribed demarcation-lines of the other; yet our semantic sense-parameters remain stable, just what they previously were. And again, we are driven to ask "why?" Why hasn't a conceptual, semantic reduction paralleled the functional reduction from one domain to the other?

It is here that one of my philosophical voices wants to speak Wittgenstein and say "Explanations come to an end somewhere."[23] But then Husserl's voice poses the question: "Why, though, is it *precisely here* that they seem always to end? Why is it that no mater how successfully we functionally reduce from one domain to another, no matter how feasible our theory-narratives of totalization are, that our intentional language never maps into our physical one? Our physicalistic language never merges with our intentional one?"

The problems and thought-experiments of this essay have been spawned, in part, from reflection on Husserl's delineation in *Ideas II* of the great ontological regions[24] and, in the same work, his abandoned attempt at a reductive totalization.[25] Husserl's attempt, in *Ideas II*, to wade into the "gray zones" of reality, where matter and spirit are intimate and mixed – in somatic and sensory animal life – and his emergence from those deep waters with a sharp delineation between material, psychic and spiritual, cannot but command philosophical respect. Unfortunately, however, while Husserl's descriptions clarify our life-worldly intuitions, they solve no problems about cross-regional reduction. Indeed, rather than resolving such problems, the clear and distinct sense-delimitation of ontological regions *causes problems* when such articulations stand beside the *fact* that causes *can* change and modify intentions, and intentions *can* change and modify causes. If the great ontological regions mark "essentially necessary distinctions" grounded in the phenomena themselves, then how is it that one such region could possibly gain priority over the other, that the other is reducible to the one? And if such is functionally and theoretically possible, what status do the regions then have? Are they mere "phenomenal" definitions useful for natural, daily life and recognized *Maya* once the reductive scientist starts work?

In *Ideas II*, we see more clearly than in any other work, the tension in Husserl's thinking that causes him to waver between a true phenomenological neutrality in relation to ontological claims, and a spiritual, mentalistic reductionism that would reduce the natural world to the spiritual world, causality to intentionality. But Husserl never published *Ideas II*. What troubled him, I think, was what is troubling us: He was deeply aware that from the *inside* of each type of science it is driven towards totalization. From the inside, each method is hermeneutically hungry, ontologically omnivorous; each has an appetite that cannot be sated until it has devoured the world. And in principle, each can succeed at its gluttonous goals. Husserl himself, still determined in *Ideas II* to procure a universal, unifying science, displayed a similar hunger. He adopted the perspective we would expect from him – the human, scientific one – and argued, feebly, that it was the totalizing champ. But Husserl was right to have remained dissatisfied with his results.

The counter-senses that we noted in section 1, articulated also by Husserl, occur at precisely the line where everyday life has established sense-boundaries that determine its capacities for coherent cognition. The epistemic-semantic problem for totalizing science is that the counter-senses become counter-sensical precisely where they do. Owing to this, *conceptually coherent* totalization from the side of either type of science gets stopped precisely where it does. While functionally, and through the eschatological eyes of its grand narrative, each type of science may continue to totalize (i.e., may have *real* effects in the Other's domain), conceptually and semantically, it remains Other to the domain it would hegemonize. When it crosses its sense-delineated boundaries and we try to fulfil an intention about what has occurred, noemata explode, we just can't comprehend – at least not in the language of the cross-over science. And since there is no neutral, higher-order language, a language beyond intentions and causes, no language that would "synthesize" them, if we wish to comprehend a cross-over event we switch to the language of the hegemonized domain. Functionally, then, and in terms of the "big stories" of science, totalization of regions seems possible; sometimes, at certain moments in the history of the sciences, it has appeared to be happening;[26] semantically and noetically, however, it leaves us unchanged, leaving only broken noemata, unfulfilled *Sinne*.

Existential agility, phenomenological calm: how to live amongst fractured totalities

We have a language of intentions and a language of causes that we cannot translate one into the other. These modes of articulation hold us fast, grip us absolutely in our understanding of different segments of the everyday world. And yet we have scientific practices that functionally burst right through the ontologico-semantic lines drawn by everyday consciousness; we have big stories, thought-experiments and methodological *apologia* that suggest that either region might encompass the other any day now; that each type of science can, in its terms, account for the hidden energies of the Other. Where, then, do we stand?

We stand where we have stood always: in the life-world. And in the life-world there have always been gaps, holes in whatever totalization we tried – religious, philosophical or scientific. We need, then, to do now what we should have done then: accept that if we seek world-totalization, ends won't meet, the center won't hold; given two competing modes of totalization, the causal and intentional ones, never shall the twain meet. The language of minds and bodies, of intentions and causes, mark modes of articulation that we can't translate across in a semantically satisfactory way. We can't even conceive *how* we could do so. Every attempt to semantically transgress the normalized sense-borders of everyday intentional life, by the use of causally reductive language, issues in counter-sense; the same holds for attempts to transgress into the causal domain with intentional language.[27] Husserl's mistake in *Ideas II* was to violate his own principle of phenomenological neutrality: his commitment to describe the world and its regions just as we find them in everyday life – without ontological commitment. But the tyranny of ontologizing reason got the best of him, too; he wanted to totalize, to make the world one, and ever so briefly he championed human scientific hegemony.

But if the sense-boundaries of everyday life remain fast *in spite of* all factual feats of reduction, in spite of all theoretically feasible inspirational narratives, then there may be only one thing to do: learn phenomenological calm, transcendental tranquility with the various languages we have for accessing the world. Simultaneously, however, as most of us already do, we must exhibit existential agility in respect to what we count real, and even when we do so. What I mean is this: we can *use* the discoveries achieved by natural or human scientific reduction without committing, semantically or ontologically, to one region's priority over the other. Indeed, this is what we *must* do if we can't translate across regions. This is not to say that the reductive and totalizing aims of each type of science should be outlawed; they should not be because these aims have provided the ideological impetus for some good and amazing things: ask anyone whose "existential *Angst*" has been assuaged by pills or whose ulcers have been cured by talk and reflection. What should be done differently, however, is suggested by Lyotard: we must lose the nostalgia for the lost narrative,[28] accept that it is *performativity* that counts, practical success that legitimates.[29] Narrativity follows behind, perhaps now, never to finish the race. And indeed, since performativity so often violates and shatters the grand unified narratives sought by Enlightenment reason, we must learn to live with and amongst incomplete wholes, micro-narratives, partial and shattered totalities – "totalities" that really *are not*. With blinders, however, each whole is and can be whole unto itself; but alas, so can its ontological Other – and *we, we* live in both at once, unblinded–ontological centaurs.[30] We epistemically sojourn in various cognitive climates, domains of intentions and causes, where never the twain seem to meet.

So is this to finally forsake the Enlightenment ideal? – the search for a grand unified theory? – for world-totalized Reason? I think it is time. But instead of saying this with negative accent, why not say that it is finally to respect the phenomena in all their diversity? – finally to bow to the world in recognition of its many, multifarious givings? – finally to accept multiple reasonings in multiple contexts with multiple meanings? Performativity may uncomfortably outdistance narrative, but does it not show that we know how to *work* with the world, "talk" to it in at least some of the various voices in which it can "hear" and respond? The fact that we cannot make all of these languages *one* that we can intertranslate and grasp does not show that we are not accessing the real. It may only mean, perhaps, that there is more to the real than meets the mind – and more, perhaps, than ever can meet it.[31]

If we are to continue in our Western, spirit, a spirit that would articulate and unleash the world in multiple discourses, forces and performative games, if we are to continue to reaffirm the Promethian proclamation, "Let the Wild Rumpus Start!,"[32] then so, too, must we learn to live with conflicting conceptual schemes, with permanently pierced noemata; we must widen and simultaneously segment our sense of the real and call into play whatever parts of it we happen to need when and where the situation demands. And never again, perhaps, shall the centaur be fooled into believing itself totally human; but then, it may no longer want or need such a belief, for it may no longer believe that anything is *totally* anything.

Notes

1. Harvey, C. W. (1989), *Husserl's Phenomenology and the Foundations of Natural Science*, Ohio University Press, Athens, Ohio, pp. 42-48.

2. The classical formulation of this thesis is given by Hume, D. (1975), *A Treatise of Human Nature*, Clarendon Press, Oxford, pp. xv-xvi.

3. Husserl, E. (1970), *The Crisis of European Sciences and Transcendental Phenomenology*, Carr, D. (trans.), Northwestern University Press, Evanston, pp. 48-53.

4. For a statement and critical evaluation of this thesis see Hempel, C.G. (1966), *Philosophy of Natural Science*, Prentice-Hall, Englewood Cliffs, N.J., p. 83.

5. Heidegger, M. (1962), *Being and Time*, Macquarrie J. & Robinson, E. (trans.), Harper & Row, New York, pp. 188-195.

6. Dewey, J. (1958), *Experience and Nature*, Dover, New York, pp. viii-ix.

7. Quoted in Hempel, p. 83.

8. Husserl, E. (1989), *Ideas II*, Rojcewicz, R. and Schuwer, A. (trans.), Kluwer, Dordrecht, pp. 231-238.

9. Husserl, E. (1970a) *Logical Investigations*, Findlay, J. N. (trans.), The Humanities Press, New York, p. 270.

10. Hume, pp. 187-218.

11. Husserl, E. (1973) *Experience and Judgment*, Churchill, J. S. and Ameriks, K. (trans.), Northwestern University Press, Evanston, pp. 72-79.

12. Husserl (1973), pp. 32-34, 72-76.

13. Husserl (1970), pp. 23-59; 111-116.

14.   Copleston, S.J., F. (1962) *A History of Philosophy: Greece & Rome*, Image Books, New York, p. 41.

15.   The most outspoken representative of this position is Bloor, D. (1976), *Knowledge and Social Imagery*, Routledge,Kegan Paul, London. But also see, Barnes, S.B., (1974), *Scientific Knowledge and Sociological Theory*, Routledge, Kegan Paul, London, and (1977), *Interests and the Growth of Knowledge*, Routledge, Kegan Paul, London.

16.   Quoted from Robinson, J.M. (1968), *An Introduction to Early Greek Philosophy,* Houghton Mifflin, Boston, p. 201.

17.   de La Mettrie, J. (1912), *Man a Machine*, Bussey, G.C. et al. (trans.), Open Court, La Salle, IL.

18.   Burtt, E.A. (1954), *The Metaphysical Foundations of Modern Physical Science*, Doubleday Anchor, Garden City, NY. See Harvey, Chs. 1-3.

19.   Pavlov, I.P. (1960), *Conditioned Reflexes*, Anrep, G.V. trans.), Dover Publications, New York, pp. 1-15.

20.   Watson, J.B. (1924), *Behaviorism*, W. W. Norton & Co., New York, pp. 1-19.

21.   Skinner, B.F. (1953), *Science and Human Behavior*, The Free Press, New York, pp. 45-58.

22.   See, e.g., La Mettrie, Pavlov, Skinner, Watson, and more recently, Wilson, E.O. (1979), *On Human Nature.* Bantam Books, New York.

23.   Wittgenstein, L. (1968), *Philosophical Investigations*, Anscombe, G.E. M. (trans.), Macmillan, New York, p. 3.

24.   Husserl, (1989), sections I, II and III.

25.   Ibid., section III, ch. iii, pp. 294-316.

26.   See the various calls to arms in the texts referenced in note 21 above. From the human scientific perspective see Bloor, D. and Barnes, S. B. referenced in note 14. And see Brown, J.R. (ed.) (1984), *Scientific Rationality: The Sociological Turn,* Reidel, Dordrecht, for further arguments and bibliographical material.

27.   For an account of the metaphysical reasons for this, as well as a description of the social and psychological obstacles encountered in any such a task, see my "The Malice of Inanimates" forthcoming in *Phenomenological Inquiry*.

28.   Lyotard, J. F. (1989), *The Postmodern Condition: A Report on Knowledge*, Bennington, G.and Massumi, B. (trans.), University of Minnesota Press, Minneapolis, p. 41.

29.   Lyotard, pp. 41-53.

30.   Ortega y Gasset, J. "Man Has No Nature," in Kaufmann, W. (ed.), (1975), *Existentialism from Dostoevsky to Sartre*, New American Library, New York, p. 154.

31.   On this theme see my forthcoming essay "Transcendental Philosophy, Plurality and Respect for the Real."

32.   Sendak, M. (1963), *Where The Wild Things Are*, Scholastic, Inc., New York.

III
ON APPLICATION: PRAXIS AND CRITIQUE

8 Coveting a Body of Knowledge: Science and the Desires of Truth

Debra B. Bergoffen

Interrogating science's desire

Science is the study of bodies: lived bodies; bodies in motion; bodies at rest; organic and inorganic bodies; celestial and earthly bodies; bodies in time; bodies in space; social bodies. As the study of bodies, it is said to produce a body of knowledge, an objective, disinterested, experimentally verifiable, coherent account of the way things are. Histories of science teach us that the study of bodies approaches the bodies it studies metaphorically as well as experimentally: the body as machine; the body as organic; the body as woman; the body as centered and de-centered; the body as the source of instinctual drives; the body as the site of disciplines. Each of these visions of the body has played a crucial role in scientific discourse. Each has privileged a different understanding of the body of knowledge to be produced. But which ever body science attends to, which ever body metaphor it appeals to, science always promises us the same thing: a reliable body of knowledge – truth.

Coming from a postmodern feminist perspective, the truth claims of science trigger my suspicions rather than my trust. But if the truth claims of science do not convince me, they fascinate me. I would like to try to understand the desires that speak in science's quest for truth. I would like to try to understand the power of these desires in/for our times.

My examination of the question: What desires speak in the scientific project? is guided by Nietzsche's re-assessment of the hierarchy of the sciences, establishing psychology as the queen of the sciences;[1] by Nietzsche's question, "Supposing truth is a woman – what then?";[2] and by science's promise to provide us with a reliable body of knowledge. Attending to this body of knowledge metaphor under Nietzsche's umbrella evokes the following question: What can science (psychology) teach us about the drive for a true (woman's) body?

To establish psychology as the queen of the sciences, and to propose that truth is a woman radically challenges our understandings of the project of science, for psychology (especially the work of Lacan and Deleuze and Guarattari) and feminist theory (especially the work of Irigaray) reveal that the body, though configured, escapes definitive configurations. A body of knowledge is not what we thought it was.

The story science tells about itself is curiously like standard accounts of evolution. Like the account traditional evolutionary theory gives of the development of life, science represents its history as a movement from the simple to the complex, from the lower to the higher, from error to truth. Like the linear progressive narrative of evolution, science represents itself as overcoming earlier naivetes and correcting past errors.[3]

In looking at the body of knowledge metaphor, my challenge to science is somewhat analogous to Stephen Jay Gould's challenge to traditional evolutionary theory. Gould, reflecting on the evidence of the Burgess Shale fossils which show many intriguing and complicated species dying out at once, suggests that extinction may be a mark of bad luck, not inferiority. According to Gould, the Shale fossils suggest that the history of life is a story of massive removals followed by differentiation amongst the remaining stocks rather than a narrative of increasing excellence.[4] By substituting the idea of desire for the notion of luck, I suggest that today's scientific theories, like today's species, are survivors and that survival is more a matter of historical contingency, than superiority. The theory that survives is the one that meets the requirements of our desire for consistency (a model of singular truth) and is contingent on the theories already in place with which it must agree, complete or extend.

Desire, history, scientific theory, truth. These thoughts guide my reading of the scientific project. Science as the drive for truth reflects the "longing for one true story" and this longing is reflected in the theories science privileges.[5] The drive for one true story demands that we choose between incompatible theories and obliterate theoretical differences. Thus, Descartes, for example, in his quest for a unified, ordered account of a hierarchal natural system turned to experiments to eliminate alternative possible theories as this elimination was crucial to his scientific ideal.[6] By putting the longing for one true story and its directive metaphors, the body of knowledge and the seamless web of knowledge, under suspicion, however, we can explore the body of knowledge metaphor for what it tells us about the desires of the scientific project and for how it might guide us toward a return of the repressed.

To bring psychology to bear on the history of science means two things: a privileging of the metaphoric and an attention to desire; for while psychology

140

respects the dominant, rational, mathematical voice of science, it hears this voice as the echo of another sounding, the unconsciousness. It listens for the unconsciousness in the meters trivialized by reason – in this case, in the off-hand, devaluated metaphors of scientific discourse. To bring psychology to bear on the history of science is to lose our innocence. We are ready to recognize the erotic dimensions of science and to examine the relationship between the sexual images and metaphors used by scientists and the scientific enterprize.[7]

When Bacon speaks of experiments as a torture which forces Nature to reveal her secrets, we pay attention to the implication of the feminine pronoun. When Bacon and Descartes link the teleology of science to the domination of nature we hear an erotic teleology. This teleology is pointed to by Ludmilla Jordanova when she sees the statue, "Nature Unveiling Herself Before Science", as indicative of the erotic relationship between the masculine viewer, science, and the beautiful woman, Nature.[8] Jordanova notes that the natural sciences have found sexuality appealing both as a subject for investigation and as a source of images, metaphors and symbols.[9] Her examination of the biomedical sciences reveals that though the body was first envisioned as a machine and later seen as organic, the metaphoric association linking body and woman, and the idea that the route to knowledge involved a looking into the body remained a constant of this scientific discourse.[10] She also notes, and this is particularly important for the project of this paper, that though the secrets of nature are identified with the secrets of women's bodies,[11] and though "science was the masculine practice of looking, analyzing, interpreting",[12] the relationship between science and its body of knowledge was ambiguous. Returning to the statue, she directs our attention to its half nakedness. Only the top is unveiled. The lower part is not.[13] The process of unveiling is not without its dangers. The desire to see, conquer, know the beautiful woman is haunted by the myth of Pandora.

Reading Jordanova's and others' accounts of the scientific project, we discover that the body science wishes to know is woman. The body it wishes to produce is Truth. If it cannot produce it all at once and now, it will settle for a narrative production. It will envision its body of knowledge as an incomplete body in the process of completing itself. But Jordanova points us to a discrepancy between the body of knowledge that is the object of science, the woman's body, and the body of knowledge that is the project of science, the masculine icon.

Supposing truth is a woman....

If we think of truth as that which science wishes to know, then attention to the metaphors of scientific discourse reveals that we have been supposing that truth is a woman for a very long time. From this perspective, what distinguishes Nietzsche is not his supposition that truth is a woman, but his awareness that this supposition is crucial to understanding our relationship to truth, scientific and otherwise. From this perspective what distinguishes Nietzsche is not his, "Supposing truth is a woman" but his "what then?" What distinguishes Nietzsche is not his supposition, but his question: What is it about the image of woman that articulates the desire for truth? From this perspective Nietzsche's "what then?" can be read in several ways. One directs us to technique: if truth is a woman how shall we seduce her? Another directs us to desire: if truth is a woman why do we want her? Both queries alert us to a certain elusiveness and danger. What is it about woman that evades us? What threat elicits the desire to control and dominate?

If, instead of thinking of truth as that object toward which science is directed, we think of it as the object/discourse produced by science, then Nietzsche's supposition that truth is a woman is disconcerting. It challenges the idea of Truth as singular in both the absolute and narrative sense and suggests that the desire for truth that fuels the productions of science contaminates the truth of the body science desires to know. From this perspective, Nietzsche's question, "Supposing truth is a woman – what then?" contains the beginnings of its own answers: we must re-think our epistemological assumptions. We must discover what it would mean if instead of feminizing the body that is the object of scientific inquiry we feminized the meaning of the body of knowledge said to be produced by science.

Psychology as the queen of the sciences: Lacan

Taking Nietzsche's directive to look to psychology for direction I take the somewhat circuitous route of turning to Lacan and his account of the mirror stage to reflect on the implications of Nietzsche's question. Here we find the human infant responding to the lure of its body image for the sake of its future and in the name of its desire. Packed into Lacan's account of the mirror stage is a theory of the distinction between the ego and the subject, a theory of desire, a theory of alienation and psychic development, a theory of the body and a theory of the power of the specular imaginary.

The grounding evidence for the mirror stage is concrete, clear and concise. The human infant's reaction to encountering an image of itself is profoundly different from the reaction of other animal infants. Whereas other animals soon grow bored with their reflected image, humans react with sustained jubilation. Unlike the chimpanzee who wants to explore the reality of the image, the infant wants to play with it. Unlike the monkey who masters the image and finds it

142

empty, the child is fascinated with the ways in which the image reduplicates its gestures. Though the infant is capable of recognizing itself in its image before the chimpanzee, it becomes captivated by its image in ways foreign to the animal.

In recognizing itself in the mirror the infant mistakes the image for itself. It misrecognizes itself. The clumsy infant is not the coordinated image. In taking the image for itself, however, the infant identifies itself with an imago, setting into play the dynamic whereby the image will determine its identity and future development. In Lacan's words:

> The mirror state is a drama whose internal thrust is precipitated from insufficiency to anticipation and which manufactures for the subject ... the succession of phantasies that extends from a fragmented body image to a form of its totality... and lastly to the assumption of the armor of an alienating identity which will mark with its rigid structure the subject's entire mental development.[14]

In mis-recognizing itself in its image, the infant alienates itself in two respects: one, it accepts the reality of the illusion of self-sufficiency/autonomy, what Lacan calls the *méconnaissance* of the ego; two, it accepts the reality of the illusion of total unity/integration.

The infant's jubilant response to its mirror image is a reflection of its desire to escape its fragmented condition. The mirror image is liberating insofar as it is recognized as reflecting the fulfilled possibilities of this desire. It is alienating to the extent that the infant/subject submits itself to the specificity of the imago, beginning the process of the ego's domination of the subject. In becoming captivated by its imago the infant/subject rejects both its fragmentation and its fluidity. It assumes its image in its fixity and thereby fixes the structure of its development.

While Lacan sees the mirror stage as a "particular case of the function of the *imago* which is to establish a relation between the organism and its reality...",[15] he notes that this stage occupies a critical place in Western culture. It is by means of the mirror image that the ego, the I, enters the place of the subject and creates the *méconnaissance* whereby the I is identified with the subject. While this alienation is a necessary part of human development, what is peculiar to Western culture, according to Lacan, is its "anti-dialectical mentality...which in order to be dominated by objectifying ends, tends to reduce all subjectifying activity to the being of the ego."[16] Lacan insists that psychoanalysts question "the objective status of the I which a historical evolution peculiar to our culture tends to confuse with the subject."[17]

This essay may be seen as engaged in that questioning. If we bring Nietzsche's question and Lacan's questioning to bear on the question of science, we discover a resemblance between the infant captivated by its body image and science's fascination with its object. This lure of the object is dramaticly portrayed in "Fat Man Little Man" the story of the Manhattan project. In this particular drama- tized slice of science history, the relationship between the scientific quest for truth, the lure of the object and power politics is strikingly clear. We see the drive to pull disparate pieces into a codified whole overwhelming all other forces. What is also clear is that the issue of truth and the question of control are inseparable. The truth sought is the theory that will control the power of the atom, harness it for military purposes. What is provocative in this account of the development of the atom bomb is the way the question of the scientists turns from, Can we know it? to Should we know it?, that is, the way the desire for knowledge becomes self-reflexive; the way science comes to question its own desire. In this movie, however, the self-reflexive moment is short circuited. The lure of the object prevails.

Like the scientists of the Manhattan project I am questioning the desire of science. Unlike them I am trying to preserve the self-reflexive moment. As Lacan suggests that the infant's *méconnaissance* is repeated in the culture of the West, I suggest that if this is so we should be able to see it in the dominant project of the West, science. Further, following the specular clue that the desire of the infant discovers itself, articulates itself and empowers itself in the imaged body, we are drawn to interrogate the body that captivates science, the (woman's) body and to question the metaphor by which science articulates the fulfillment of its desire to know what captivates it, the body of knowledge. We are driven to the following question: when science envisions its feminine body (its object) as a body of knowledge (its product) is it replaying the drama of the mirror whereby the infant substitutes an image of its body for its experience of embodiment? Is this re-enactment an institutionalization of a primordial *méconnaissance*?

I am tempted to think that it is. I am tempted to suggest that the body of knowledge metaphor calls upon the promise of the mirrored body image. I am tempted to suggest that this promise of the body image, now imaged as the body of knowledge, grounds and empowers the project of modern science: to translate everything into the coherent, the measurable, and the calculable. But I am also tempted to suggest that attending to the tension between the metaphoric representations of the object and project of science allows us to discover the powers of the desires of science and to use this discovery to challenge current configurations of the scientific project.

Where do these temptations lead? To a confirmation of Lacan's observation. To some insights into the dynamics of this cultural *méconnaissance*. And, to some suggestions for alternatives to our alienation. If we attend to Lacan's

144

observation that the body image is simultaneously essential for psychic development and a *méconnaissance*, we discover that in designating itself as a body of knowledge directed toward a woman's body, science opens itself to the possibility of recognizing its *méconnaissance*.

Eve's body

The feminine body plays in several registers. So far only one of these registers has been clearly heard. So far the register which configures the desire of/for the object according the designs of mastery has out sung the others. Here the mysterious, but passive body waiting to be unveiled, has held out the promise of order. The truth of this register of the body is the truth of reason: coherent, contained, controlled being. So far the lure of the body metaphor has been the lure of the fulfillment of the desires of reason/the ego.

Lately, in the work of Irigaray for example, the idea of the passive woman's body has been challenged. Here, the anxieties of unveiling alluded to in the statue of "Nature Unveiling Herself Before Science" are articulated. Perhaps the woman's body is not amenable to the structures of reason. Perhaps the revelation of this body's mysteries will threaten the body building project of science. Here the woman body metaphor moves against its original intentions/intuitions. Instead of reinforcing the desires of reason it disrupts them.

This disruption, however, is not the work of some alien power. It is part of the work of the metaphor itself. Woman, the body in waiting, the passive body waiting to be known, is also the body of temptation, the body of Eve bringing a knowledge that threatens to undo the given order of things. Science, in articulating its desired object according to the image of the woman's body has put an image with metaphoric powers into play. Modern science called upon the metaphor's powers to empower an image of the desires of reason. A science directed by Nietzsche's questions calls upon the metaphor to image other desires. A new balance of power emerges from within the powers of the ambiguities of the metaphor. The desire for the complete, controlled and controllable body comes under scrutiny. The desires of disruption demand attention.

If the truth of the experienced body is repressed by the image of the visualized body; if this repression expresses the desire for order, coherence and rationality; if this desire succeeds in passing the imaged body off as the true body; then is the truth of the body (what now passes for) its untruth? Is the truth of the body the lie of its status as a stable image and the power of its being as mobile metaphor? Is the truth of the body other than the truth of its image as a delineated terrain with discrete boundaries that function according to discernable

145

laws? Is the truth of the body the truth of the desires that play at the edges of the reflected imago?

These questions, this attention to edges, draws us toward new body metaphors and focuses our attention on the desires at work in the productions of reason. In the work of Deleuze and Guatarri it transforms the machine body metaphor from a way of envisioning the interconnected, reasonable, orderly and self-consistent workings of nature to a way of insisting on the uncoded, coded, overcoded and decoded productions of desire.[18] In the work of Bataille it provokes an attention to the limits of reason. Without disputing the legitimacy of science, Bataille confines science to the domain of reason and subordinates reason to the domain of chance. He writes,"...in the long run the course of the world obeys law. And since we're rational we see this; but the course of things escapes us at the extremes."[19] Chance is at the extremes. Science, as Bataille sees it, is part of our desire to confer necessity on ourselves and the world; part of our desire to avoid chance.[20] Expressing another desire, the desire of "the giddy seductiveness of chance"[21] Bataille counters the traditional ideal of science, the ideal of totality, with another ideal, anti-totality. Instead of pursuing the body imago, Bataille would subvert it. For him, "Knowledge destroys fixed notions and this continuing destruction is its greatness or more precisely its truth."[22]

If we see the difference between a live and a dead metaphor as the difference between a disruption of boundaries that provokes a new way of seeing and a boarder crossing made banal by repetition then we might see this project of suspicion as a project of metaphoric re-vitalization. Instead of accepting nature as feminine as a matter of course we see this as a metaphoric move and ask what it signifies. Instead of accepting the phrase body of knowledge as a dead metaphor we seek to revive it.

Two things are striking about the body of knowledge metaphor: first, it aligns knowledge with the image of a totalized body; second, it aligns reason with its other. The metaphor disrupts the mind-body dualisms of the Western tradition. Against the idea of the body as the other of knowledge, the metaphor transforms the body into the sign of knowledge. It disrupts the boundaries of mind and matter in order to assure us that matter is amenable to mind. The point of the metaphoric disruption seems to be one of containment. Body does not escape the boundaries of reason. It images them. That the metaphor expresses a desire similar to the desires that empower the mirror stage seems to be indicated by the particular body invoked by the scientists – the woman's body. The most unruly, the most secretive, the most beautiful and seductive, that is, the most disruptive body; the body most unlike rational discourse, this is the body that will be taken up by/into the body of knowledge. Reason, science, will speak its truth. Though the body carries the sense of being the other of reason, hence the need for the mind to control it when it was envisioned as a machine and the need to tame it when it was envisioned as an organic wilderness, the body's otherness can now, under the guidance of the metaphor, be thematized as superficial. As the superficial other, the body can be appended to reason in the name of an ideal of Truth that represses all thought of *méconnaissance*.

But, supposing we give the metaphor back its mobility by feminizing it through and through. Suppose we ask it to suggest other relationships between the body and truth. Suppose we bracket the assumption that reason is the truth of the feminine body and ask whether the metaphor might direct us to another thought, that the feminine body is the truth of reason. Suppose, instead of being led by the thought of the body as other to the thought of the body as the object of reason, we follow the thought of the body as other to the thought of the body as the desire of reason and follow the thought of this desire beyond reason. Could there still be a singular, true body of knowledge? Would we still be seduced by the imago of totalization? Supposing Truth is a woman...what then?

Pursuing Nietzsche's question under the influence of Lacan provides a perspective from which we can suggest that the body of knowledge metaphor has been functioning in the register of the imaginary. The move by which it allows body and knowledge a point of intersection is fixed/fixated on the image of the coherent body. The meanings of this body reflect the desires of the *méconnaissance*.

By remembering that the body represented by the body of knowledge called science is a woman's body, however, we create a place from which the imaginary freezings of the metaphor are challenged. Supposing truth is a woman? Supposing the woman's body, not the body of knowledge is truth? Supposing the body of knowledge as feminine alerts us to the complexities of the desires for truth? Supposing these desires are and are not satisfied by the produced body of knowledge? Supposing this is the point of figuring the body of knowledge as woman?

The metaphor and its ursurption by the imaginary alerts us to the play of desire. There is the desire of the imaginary, the desire of *méconnaissance*, but there are also the desires of the metaphoric, the desires of the play of the signifier. The history of the body of knowledge metaphor reminds us that the imaginary and metaphoric registers intersect. As the desires of the imaginary can empower the draw of the metaphor, the desires of the metaphor can challenge the pull of the imaginary. Truth as woman's body undermines the seductive powers of the imagos of totalization, and this, it seems to me, is what lies at the heart of Nietzsche's question, Supposing truth is a woman....?

Re-awakening the metaphor allows us to challenge its original intentions without challenging its original boundary crossings. It allows us to affirm the insight that empowers the metaphor, that the body and truth intersect each other, and to reconsider the *méconnaissance* that hardened the metaphor. Instead of bi-furcating the body into an object, feminine body,[23] and a product, truth (masculine?) body, where the object is taken up by the product, the body itself, as metaphor, interrupts this object-product relationship. Where we first supposed that woman's body was the object of science, now we suppose that it is the object and project of science. Woman's body is not merely the metaphor for the object to be described by science, but the metaphor for scientific discourse itself. Led by this metaphor, science becomes the speech that recognizes its other as other. It produces a body that questions the body produced. The woman's body, originally the carnal body that had to be controlled by science now becomes the carnival celebrated by science.

Where experiments were originally called upon to establish the unity of Nature; where their original function was to eliminate the "and" in the name of the "is"; where their first mission was to subordinate the multiple to the one; they are now called upon to empower the empiricism of the "and".[24] Further, under the impact of the indetermination thesis, the relationship between experiment and hypothesis must be rethought. As we experiment in the name of the heterogeneity of the given, the theories we devise to account for the given can never claim to exhaust the possibilities of the given. Where experiments were originally seen as the way to move theories from the domain of hypothesis to the realm of fact, truth, they may now be seen as ways of reminding us of the inconclusiveness of our theories, truth.

The question mark, the empiricism of the "and", the indetermination thesis, the attention to heterogeneity are not the other of science. Returning to the scientists we note the way the value of the margin is recognized in recent evolutionary theory. Differing somewhat from Darwin's view, contemporary theorists explain the phenomenon of speciation (the emergence of new species) in relation to the phenomenon of homogenization. It is now believed that new species branch off from the parent stock at the geographical fringes where homogenizing influences are weaker.[25] Shall we suggest that this vision of nature speaks from the desires of the edges? Shall we ask science to encourage the speciation that enriches speculative life by challenging its own homogenizing values/tendencies? Shall we demand that the body of knowledge metaphor once used to validate the values of homogeneity now be called upon to combat them? Shall we suggest that the question mark replace the period as the sign of science? Shall we propose that Nietzsche's "Supposing truth is a woman – what then?" be taken as the model for the structure of scientific discourse: a discourse that begins by recognizing itself as hypothetical, that pauses in its reflection and that questions the meanings of its affirmations?

As the infant cannot develop without the guidance of its body imago, a culture cannot function without a body of knowledge. As the body imago situates the infant within its world, a body of knowledge gives a culture its world. Noting

148

the *méconnaissance* and alienating powers of the imago we cannot reduce the imago to a false vision. We can, however, be alert to the ways the image refuses the metaphoric function. Recognizing the power of the image we can refuse to be dominated by it.

Which means? Which means continuous attention to the dynamic relationship between order, disorder and desire. Which means? Which means an end to the domination of science by the desire for order; an end to the domination of the culture by the homoginizing demands of scientific discourse; and a valued attentiveness to the metaphoric voices of a body of knowledge envisioned as woman.[26]

The body of knowledge metaphor permeated by feminist and psychoanalytic insights and punctuated by Nietzsche's question, Supposing truth is a woman? pushes the experimental skepticism of the scientific project to its limit. Now skepticism, instead of being the means to the scientific end of truth (e.g., Descartes) comes to be understood as the end itself.

Identified as the telos of science, skepticism moves science to understand its production of a body of knowledge as expressive of the desire for continuity and coherence and cultivates a suspicion of this desire. This suspicion allows it to reconsider the value it accords to truth as singular and to distinguish itself as that body of knowledge which critiques/undermines the coagulations of the drive for order. From this perspective, science is that body of knowledge which saves the world from being lost in the word. It is the discipline which saves experience from disappearing into that interpretation of it that succeeds in getting itself called truth. The point of this perspective is to preserve the radical empiricism of science insofar as empiricism is taken to be that position which evades the imaginary desires for certainty, coherence and consistency by being open to the interruptions of the Real.

References

Bachelard, G. (1970), "Epistemology and History of the Sciences," in Kockelmans J.J. and Kisiel T.J. (eds.), *Phenomenology and the Natural Sciences*, Northwestern University Press, Evanston, Il.

Bataille, G. (1988), *Guilty*, Boone, B. (trans.), The Lapis Press, Venice.

Butterfield, H. (1957), *The Origins of Modern Science*, The Free Press, New York.

Deleuze, G. and Parnet, C. (1987), *Dialogues*, Tomlinson, H. and Habberjam B. (trans.), Columbia University Press, New York.

Deleuze, G. and Guattari, F. (1983), *Anti-Oedipus*, Hurley, S. and Lane, H. (trans.), University of Minnesota Press, Minneapolis.

Gould, S.J. (1989), *Wonderful Life: The Burgess Shale and The Nature of History*, W.W. Norton, New York.
Harding, S. (1989), "Feminist Justificatory Strategies," in Garry, A. and Pearsall, M. (eds.), *Women, Knowledge and Reality*, Unwin Hyman, Boston.
Haraway, D. (1986), "Primatology is Politics by Other Means," *Feminist Approaches to Science*, Bleier, R. (ed.), Pergamon Press, NewYork.
Jordanova, L. (1989), *Sexual Visions:Images of Gender in Science and Medicine Between the Eithteenth and Twentieth Centuries*, University of Wisconsin Press, Madison, Wisconsin.
Lacan, J. (1977), *Ecrits*, Sheridan, A. (trans.), W.W. Norton, New York.
Nietzsche, F. (1966), *Beyond Good and Evil*, Kaufmann, W. (trans.), Vintage Books, New York.
Snyder, P. (1978), *Toward One Science*, St. Martins Press, New York.

Notes

1.   Nietzsche's proposal does not go against the grain of the history of science. As biology was recognized as the most significant science of the Renaissance, and physics was established as the first science of the modern age, psychology would become the model of science in a post modern world. See Butterfield, p. 49.

2.   Nietzsche, p. 2

3.   For a defense of the idea of progressive, scientific truth from a phenemonological perspective see Bachelard, pp. 317-352.

4.   See Gould.

5.   Harding, pp. 189-202.

6.   Butterfield, p. 125.

7.   See for example: Garry, A. and Pearsall, M. (eds.), (1989), *Women, Knowledge and Reality*, Unwin Hyman, Boston; Tuana, N. (ed.), (1989) *Feminism and Science*, Indiana University Press, Bloomington; Jaggar, A. and Bordo, S. (eds.), (1989), *Gender/Body/Knowledge*, Rutgers University Press, New Brunswick; and Gallagher C. and Laqueur, T., (1987), *The Making of the Modern Body*, Berkeley, University of California Press, Berkeley.

8.   Jordanova, pp. 87-89.

9.   Ibid., p. 19.

10.   Ibid., p. 50.

11.   Ibid., p. 96.

12.   Ibid., p. 110.

13.   Ibid., p. 5. Salvador Dali's "The Woman Aflame" also plays on this theme. According to the 1980 exhibition catalogue, "The beautiful faceless woman symbolizes all women. Her drawers hide her secrets and contain her mysteries."

14.   Lacan, "The Mirror Stage as Formative of the Function of the I," p. 4.

15.   Ibid.

16.   Lacan, "Aggressivity in Psychoanalysis," p. 23.

17.   Ibid.

18.   See Deleuze and Guattari.

19.   Bataille, p. 71.

20.   Ibid., p. 74.

21.   Ibid., p. 72.

22.   Ibid., p. 25.

23.   Haraway (1986), p. 82.

24.   Deleuze and Parnet, p. 57.

25.   Snyder, p. 28.

26.   How such a body might speak or be spoken is suggested by Luce Irigaray, "Is the Subject of Science Sexed?", Tuana (ed.), pp. 58-68.

9 Postmodernism and Medicine: Discourse and the Limits of Practice

Chip Colwell

...this projection of illness onto the plane of absolute visibility gives medical experience an opaque base beyond which it can no longer go. That which is not on the scale of the gaze falls outside the domain of all possible knowledge.[1]

...health replaces salvation...because medicine offers modern man the obstinate, yet reassuring face of his finitude; in it, death is endlessly repeated, but it is also exorcised; and although it ceaselessly reminds man of the limit that he bears within him, it also speaks to him of that technical world that is the armed, positive, full form of his finitude.

– Foucault (198)

While the issues and arguments that are usually associated with postmodernism are obviously relevant to the concrete experience of our daily lives the immediate relevance of the issues raised by a postmodern philosophy of (physical) science seem at first to be a bit distant. Issues regarding our identities and desires are crucial to how we understand ourselves and others and how we go about expressing that understanding in our social and politcal interactions whereas issues regarding whether muons are real or theoretical entites have little if any direct impact. But there is at least one science which has a direct impact on our lives. Medicine is the science in which *we* are the objects of knowledge and of scientific intervention. As such medicine has both a theoretical and a practical impact on our lives. On the one hand, it is through medicine that the theoretical structures of our understanding of ourselves (at least as physical beings) are produced. On the other hand, and perhaps more importantly, it is through medicine that the possibility of the continuance of life and health in the face of the many insults to which we are prey is made possible. As such I want to ask the question of postmodernism and medicine in the most pointed way possible. Does one want a postmodern doctor?

We need to pose this question on two registers since we encounter, and confront, medicine in two ways. The more obvious one is the relation we enter when we go to the doctor or the hospital, in the attempt to either maintain or restore health. But we also confront medicine in the marketplace, in its public functions with regard to the community at large. In order to address this question I shall examine the functioning of medicine in these two locales in an attempt to distinguish the features of a postmodern physician from that of a modernist physician. Following that I shall address our confrontation with medicine from the other side and ask the correlative question: does one want to be a postmodern patient?

But what is, or would be, a postmodern physician? Unless we are to beg the question we must first ask what postmodernism is, what its definition is. This of course is a difficult question to ask, or rather to answer, since at least one of the characteristics of postmodernism seems to be the refusal of the possibility of meaning, of definite definitions. The "paradigmatic" definition of postmodernism is that of Jean-François Lyotard.[2]

Postmodernism is the standpoint from which we are incredulous towards all grand meta-narratives. That is, it is the refusal to accept the possibility that there is any discourse in which the whole story can be told and to which all other stories (narratives, claims to truth, sciences, philosophies, etc.) can be referred in order to determine their credibility. It is, as our analytic colleagues would say, a non-foundationalist position.

We, of course, must next ask about the relationship between medicine and narratives. This is a more complex question than the previous one (since the previous question cannot be answered, it is, in one sense, rather easy to answer it). Now biology, organic chemistry, anatomy, physiology and the other sciences that operate within medicine are all, of course, narratives (at least in, or from the standpoint of, postmodernism). But medicine is itself a certain narrative or, in Michel Foucault's terms, a certain discourse, a certain discursive functioning that generates its objects, its modalities, its concepts, its strategies. This is not to say that medicine is merely a discourse. Rather, medicine is a form of discourse/practice, a way of seeing and a way of acting that is structured by regimes of enunciating, as Foucault shows in *The Birth of the Clinic: An Archaeology of Medical Perception*. What I shall do below is to adumbrate Foucault's argument regarding the discursive functioning of medicine and then go on to look at the way in which medicine functions presently. I shall then

154

return to the question with which we began.

While *The Birth of the Clinic* was written in Foucault's more structuralist period, a period in which he emphasized discourse over practice, we must recognize that medicine is very much a discourse/practice. Foucault has three stories to tell here, or at least three meta-stories since the book is filled with particular stories. The first is the genealogical story of the multiplicitous lines of descent that form the matrix out of which anatomo-clinical medicine emerges. The second is the archaeological story of that matrix, of the "historical and concrete *a priori* of the modern medical gaze." (Foucault, 192) The third is the story of how the modern concept of finitude functions in the development of modern medicine, of how death becomes the standpoint from which life is understood and from which the individual becomes an object for science.

As to the first story while it is generally argued that Foucault's genealogical period begins after *The Archaeology of Knowledge*[3] anyone who now returns to *The Birth of the Clinic* cannot fail to recognize its genealogical nature.[4] If genealogy is the tracing of the multiplicity of lines of descent of a current edifice of knowledge then this is exactly what is done in this work. Foucault shows that modern medicine is not born from a return to a pure empiricism of the phenomena of disease. Instead it emerges from a matrix that is formed by economic factors, changes in the structure and relative importance in the social field of the hospital, changes in the status of the physician and the patient, changes in the conception and operation of public assistance, among others. It is these changes which establish the space in which modern medicine arises.

We can see this in Foucault's analysis of the shift in the relation between the hospital and medicine. While this transformation occurs at the intersection of a number of lines of descent a glance at two of these, the patient and the hospital, will suffice for our purposes here. In the classical age (roughly, the 17th and 18th centuries) 'classificatory' medicine studies, classifies, *knows*, diseases in their relation to each other; their relation to the patient is a secondary consideration. (Foucault, 5ff) Within this system the status of the doctor, the patient and the hospital are related in a manner that appears peculiar to the contemporary eye. The hospital is not the privileged site of disease it is now, indeed it is considered as the last place that one can see and know the disease itself. (Foucault, 17) This is because the hospital is not the 'natural' site of disease. Disease arises in the home, in the community, and it is there that disease takes its natural course. As such, it is there that disease must be studied, diagnosed and treated. For classificatory medicine both the doctor and the patient are lost in the hospital, the patient because s/he loses the natural framework of care that is found in h/er family and community, the doctor because the hospital is where diseases intermingle and disguise themselves from medical knowledge by becoming associated with and exhibiting symptoms that are not natural to them. Moreover, the hospital places a double economic burden on society–the patient is not only unproductive, if hospitalized s/he also has to be kept at the expense of the state. (Foucault, 18)

But the hospital became, and remains, the privileged site for the detection and treatment of disease. In part this is due to a shift in the social/political constitution of the patient. The sick individual comes to be seen as one who deserves public assistance, a responsibility of, rather than a liability to, the government and society in general. (Foucault, 32) Moreover, as the use of clinical training of physicians increases, the patient who offers h/er body and disease to the hospital comes to be seen as an asset. S/he becomes the one whose body becomes the locus of the production of knowledge, by being both the tool used to train new physicians and the object of medical experimentation, and thus a contributor to the common good rather than a drain on it. (Foucault, 84-5) This change in the social/political status of the hospital intersects with a transformation in the object of the medical 'gaze.' In classificatory medicine the 'gaze' is trained on the disease-in-itself, a disease that can be masked when it intermingles with other diseases, when it is tainted by symptoms other than those that are specific to the disease. But when medicine begins to train its eye solely on symptoms, when disease becomes nothing more than symptoms (I will return to this point below), the hospital becomes the privileged locus of disease because it is there that symptoms can be observed with the greatest efficiency and perspicacity (precisely because it is reconfigured as a panopticon in which symptoms cannot hide). At this point it becomes possible to constitute the hospital as a "*neutral* domain" in which disease can reveal itself. (Foucault, 109) This shift in the locus and direction of the medical 'gaze' from the 'natural' site of disease (and the patient) to the construction of a 'neutral' site for disease (and the patient) is a crucial one for understanding the functioning of current medicine. This is because it is this 'neutral', objective, space which, in part, establishes the boundaries of the discourses which translate patients into medical objects and generate medical discourse as a standard of truth, a 'grand narrative.'

The second story is an archaeological one. Here Foucault unearths the shards and remnants of discourses that contemporary medicine has left behind, the discourse of the classificatory formation of medicine, the discourse of the 18th Century clinic with its signs and symptoms, the initial discourse of tissues of anatomico-clinical medicine, and the outset of the discourse that is our own, the clinical discourse that finally relinquishes the ontology of disease for causal discourse of the relation between the body and the agent which causes disease, a relation that is always particular, that is to say, individual. Let me adumbrate this by comparing the difference between classificatory and clinical medicine regarding the relation of symptom to disease.

Classificatory medicine distinguished between symptoms and signs.

(Foucault, 90) The symptom (e.g., cough, fever) is what allows the physician to diagnose the disease; it is the form by which the disease announces its presence. The sign (e.g., pulse, cyanosis) is what informs the doctor of the course of the disease; it is a prognostic sign. Both function as signifiers for the disease which remains both 'visible and invisible,' visible in that it shows itself through the symptom, invisible in that it is never the disease itself that is seen. The key points here are the existence of the disease as an entity beyond the symptoms and the direct, fixed link between symptom and disease. For clinical medicine the distinction dissolves so that the symptoms *are* the disease, or rather, the disease is nothing more than its manifestations, its symptomatology. (Foucault, 91-6) The relation between signifier and signified collapses in that the signifier signifies itself. It is at this point that "the essence of the signified – the heart of the disease – [is] entirely exhausted in the intelligible syntax of the signifier." (Foucault, 91) What had remained invisible to classificatory medicine vanishes leaving only the visible.

This may seem at odds with the functioning of current medicine in that we tend to think of the symptoms of, e.g., *pneumocystis carinii pneumonia* as shortness of breath, fever, productive cough, etc., while we think of the diagnosis as based on a positive culture for that particular organism and, as such, the disease as being the overrunning of the body by this organism, or simply as this organism. But if I have understood Foucault correctly that positive culture, that overrunning of the body by the organism, is itself a symptom rather than something over and above the other symptoms of the disease. And, indeed, it is here that the force of the word 'syntax' is felt for diagnoses are based not on single symptoms but on series of multiplicitous and aleatory symptoms. That is, diagnosis and treatment are based on series of symptoms (indeed, series of series of symptoms), any number of which are associated with other diseases, and which are aleatory with respect to the particularity of the patient which has the disease. It is at this point that the 'uncertainty' of classificatory medical knowledge is transformed into the probabilistic nature of clinical medical knowledge. (Foucault, 97) Rather than seek the absolute truth of disease the frequency and repetition of symptoms are tabulated and related to one another with increasing degrees of probability to form diagnoses. Now, we need neither absolute truth nor ontology to deal with disease.[5]

Lastly, there is the discourse of death. Death, finitude, is transformed from that which is the limit of life, the moment of cessation of life that hides life's secrets forever, to the vantage point from which life is finally understood. It is that which finally allows the 'gaze' of the physician to penetrate the opaque surface of the skin, via the autopsy, and see what both life and disease have left in their wake. For clinical medicine, knowledge of the living is based on knowledge of the dead. (Foucault, 171) The interior of the living remains

157

invisible, it hides itself from knowledge. It is only through the medium of death that knowledge is possible.

As Foucault points out, the matrix of clinical medicine does not come to be organized around the corpse because of a sudden availability of bodies in the post-Enlightenment era. (Foucault, 124ff) Neither religious nor moral scruples had hindered the availability of corpses for some time before clinical medicine recognized the usefulness of the autopsy. Instead, it was a shift in the matrix of medical knowledge, its historical *a priori*, that constituted the corpse as a possible object of knowledge. This shift, broadly stated, occurs when symptoms are disconnected from disease itself and constituted as reactions of the body to a disease causing agent, i.e., as part of the disease rather than as merely a signifier of disease. It is at this point that medical knowledge is organized in such a way as to produce the dead body as something worthy of its 'gaze.'

But what does this tell us of modern medicine? Let's run through the three stories again. Each has something to tell us, or perhaps warn us about. The genealogic story warns us that medical knowledge does not have a direct, unmediated relation to life, the body, disease. It is always mediated by the institutional structures where it arises, the social constraints of the culture in which it arises, the political maneuvers amongst the producers of knowledge, the political structures that fund research (and treatment), by an entire array of contingent effects that form the site(s) where knowledge is produced. The warning here is that the foundation of medicine is not a bedrock of absolute knowledge but a site that is itself contingent. In Lyotard's terms it is one narrative among others. But in Foucault's terms we can say more than that. Medicine is a discourse that does not arise from a privileged relation to an object that precedes it; instead it is a discourse that arises at the intersection of other discourses, that is generated by other discourses, and which in turn generates its objects.

The archaeologic story contains a similar warning. Here the warning is that the collective consciousness of medical knowledge does not have an unmediated relation to life and disease. What counts as medical knowledge, what counts as having truth value, is not something that lies within the body. The *body* does not speak to doctors in its own language. What speaks is an entity constituted within a discursive formation that is designated by the term 'medical knowledge,' a discursive formation that surrounds the body and lends its own way of speaking to the body. We can see this in the operation of the various technological devices (ECG, EEG, phono-echo-cardiogram, ultrasound, CAT scan, angiogram) that wrest a certain speech from the body, a speech that is always already inscribed in the discourse of medicine. These "inscription devices" transform the body into "written documents" that obey the rules of medical discourse precisely because these devices are themselves products of that discourse.[6] What *patients* say to doctors also has to be translated into this discourse in order for their utterances to become statements that will function effectively in this realm.

The story about death becoming the vantage point from which life is understood has a somewhat different warning, one I suspect we are less likely to want to hear. That warning is about the difference between medical 'knowledge' and the epistemic relation between the doctor and the patient, or what it is that the doctor may 'know' about the patient. Once the body has died it is possible to achieve knowledge about it. This for two reasons. The more obvious reason is that it becomes possible to fully examine the body without concern for disrupting the processes of life. All parts of the body may be opened up, examined, tested, dissected. The second reason is simply that it is dead. It has stopped living, stopped changing, stopped reacting, stopped exhibiting its own idiosyncratic responses to the disease process, stopped picking up secondary infections, stopped having side effects to treatment, stopped refusing 'necessary' procedures, stopped being a becoming. The reason that doctors can only give probabilistic diagnoses and prognoses (besides the fact that they sometimes fail to diagnose and treat as well as they might) is that each body has its own relation and reaction to disease and treatment and it is impossible to apply all the means of producing knowledge at medicine's disposal to a living body (and have it continue to be a living body).

To be sure, modern medical technology has striven to overcome this obstacle. It has developed technologies that peer inside the body (bronchoscopy, colonoscopy, cystoscopy, gastroscopy, laparoscopy); such devices allow the entry of the doctor's eye into some of the body's cavities. It has developed the method of biopsy by which a small amount of tissue is removed from an organ (not enough to alter its functioning) in order to perform a mini-autopsy on it. Radiology and other scanning devices allow a less direct vision of the interior. But it is precisely the point that these technologies are developed in order to overcome the limitations on medical knowledge placed there by the presence of life and generated by the structure of the current order of its discourse/practice.[7]

Without attempting to characterize the postmodern (or modern) physician (for the moment) we can see what the central concern is for those who would pose the question raised at the beginning of this essay to themselves. What is worrisome here is the thought of being treated as a merely discursive object. We fear the loss of the reality of our pain, our discomfort, the disruption in our lives, our disease under the gaze of one who simply translates all this into mere words that cannot communicate our experience. As Elaine Scarry has pointed out, one the problems of the discursive structure of the contemporary world is the lack of a finely differential and expressive 'ordinary' language for pain.[8] What we lack on Scarry's analysis is an ability to express pain in such a way as to communicate it to others. In medicine this is problematic because while descriptions of pain are used for diagnostic purposes the lack of the ability to express that pain in terms of the human experience of it distances the physician from, as it were, the pain in-itself, i.e., from the patient as the bearer of pain.[9] Moreover, the medical discourse of pain is one developed by medicine for patients,[10] and one that is taught to patients to enable them to communicate in that discourse. What we fear is the loss of the tangible reality that we experience ourselves as, the dispersion of that reality into discourse under the gaze of that discipline that we turn to in order to continue that reality.

This is analogous to the other fear that the invocation of the term 'postmodernism' evokes in some. What is feared is the loss of the reality of one's being, the reality of the experience of one's self, one's identity, in the denial that there is an originary subject that exists prior to the constitution of one's identity in the swirl of the multiplicitous discourses that construct the subject positions through which individuals move. This, I think, is the core of most of the criticism of Foucault. To turn to the issues raised by *Discipline and Punishment* and *The History of Sexuality: Vol. I*[11] for a moment, Foucault's argument here is that our subjectivities are wholly constituted by the discourse/practices of the power/knowledge structures that we find ourselves within. The problem that social/political theorists tend to see here is that such a notion of the subject leaves no room for liberation, for liberative strategies, for political struggles. But the problem runs deeper than that for it questions the very worth of our own selves, the parameters of our relations to our selves. If we are nothing but constructions of discourse/practices then there seems to be no room for notions of self worth, dignity, responsibility, pride in accomplishments, all those notions which have traditionally formed our experience of our selves. In short, what is lost is our experience of ourselves as (at least potentially) self-determining human beings.

Correlatively, again, what we might well fear losing under the gaze of a postmodern physician is our experience of the reality of our lives, our bodies, our pain, the limitation of our activity, in the construal of it as merely discourse. Now we must realize that Foucault never denies the reality of the subject or the body. For him the subject is no less real for being a product of discourse/practice.[12] Correlatively, to view the body, disease, and the human experience of both as a construct of discourse/practice is not to deny the reality of that experience. Nor is it to deny the reality of either disease or the body. It is, rather, to argue that this reality is an effect rather than an irreducible object. And insofar as it is an effect its reality is contingent and quite possibly temporary.

Let's consider the discursive construction of the subject in its identity as a *patient* for a moment or two. Anyone familiar with the functioning of present day hospitals is painfully aware of the discursive constitution of the patient. After all, you can't tell a patient without a chart. While the preceding statement may seem facetious it is quite true. Once past the initial stages of hospitalization, during which the foundations of the chart are laid down, it becomes nearly impossible to do anything for, or with, the patient without the chart in hand. This is much more than the question of identifying the patient through h/er ID number or verifying a method of payment. It is only through the chart that the patient may be approached as an object for the medical 'gaze.'

Once one is past the administrative nightmare of name, address, emergency contact and method of payment (information documented on what is called the 'face sheet'!) the discursive constitution of the patient is initiated by the History and Physical. This begins with the story that the patient tells the doctor about what brought h/er to the hospital, a story of pain, swelling, shortness of breath, etc., a story which the physician writes down in as brief a manner as is possible (it is rare to see one that is longer than two thirds of a page). It continues on through a past medical history, family medical history, a social history (employment, marital status, drug, tobacco, alcohol use), and into a physical exam. Here the examination of the body via visualization, palpation and auscultation is reduced to a discursive description (sometimes containing stick figure drawings to indicate neurological or circulation deficits by region of the body). Lastly, we find an assessment, or diagnosis, and a plan of action/intervention.[13]

The chart continues with physician's orders (often entered into a computer system with hard copies placed in the chart only every 24 hours),[14] progress notes detailing the course of the disease and treatment, clinical records of vital signs (blood pressure, pulse, temperature, etc.), consultation reports from specialists, lab data, reports from the various diagnostic services (x-ray, ECG, EEG, Nuclear Medicine), surgical reports, respiratory therapy records, social service notes, consent forms, etc. While photographs occasionally accompany these reports there is always a discourse that accompanies and explains these visual documents, that turns the images into objects of knowledge. Vision explains nothing until it is rendered into discourse.

Moreover, this vision is one that is always informed, formed, trained (disciplined?) by discourse. Anyone who has looked at an X-ray, or an ECG, or a CAT scan, knows that these images or inscriptions refrain from explaining themselves. Most of us remember the difficulty we had when we first peered into a microscope and were unable to identify or even see any of those things that we had to in order to pass lab. In each case a certain training is required; one has to be trained to see what is there, to differentiate an initially amorphous series of gradations of light, or squiggling lines, into discrete and expressive inscriptions. Such training is done, to a large extent, discursively. The objection might be raised at this point that it is a question not of discursively forming vision but, as Wittgenstein would say, of pointing out. But this is not the case. Reading an X-ray is not merely a question of recognizing certain features, it is a matter of *saying* what these features are, what they mean, what state of health or disease they represent. Furthermore, it is a matter of discursively integrating these images into diagnostic, prognostic and prescriptive regimens.

The point here is that to enter the hospital, the clinic, the doctor's office is to hand oneself over to a discursive regime that is (usually) not one's own, that one is (usually) unskilled in, and that (re)constitutes one's body and life in a manner over which one has little control. It is to be turned into an object of knowledge. The question is, of course, just what sort of object this is, what sort of knowledge this is. It is here that we can locate the question of just what would characterize a postmodern physician, what the difference would be between a postmodern and a modern doctor. The postmodern doctor is, as I have already noted, a non-foundationalist. What that means is that she recognizes that the discourse that constitutes an individual as a patient is only one discourse among other discourses, specifically only one of the many discourses that constitute the individual.

I will return to the characterization of the postmodern physician below. That characterization will depend on its differentiation from the modernist physician (and necessarily so as postmodernity is a modification of modernity). The modernist physician takes science in general and medicine in particular as the grand meta-narrative, the discourse of discourses, that reveals the truth of life. He, of course, would not construe medicine or science (or life) as (mere) discourse. The discourse and all its accompanying technological devices are simply ways of representing the truth of life, of the body, aids to the medical gaze in which the body is reflected. This not to say that modernist doctors do not recognize that there are other aspects to life. But it is to say that these other aspects are just that, merely aspects, possible contributing factors to the disease, aids to diagnoses, factors that determine the specific instructions to be given to the patient on what s/he can or cannot do. Medicine remains the grand narrative that knows the truth of that which allows the possibility of all those other aspects. From this standpoint, it is only when the body is submitted to the truth of medicine that it will continue to live and function as the condition of the possibility of those other aspects. Since modernist medicine sees itself as knowing the truth of that condition, and moreover that which maintains the existence of that condition, it takes itself as the foundation of the foundation of life.

Moreover, in the modernist form of medicine the object of knowledge that the body is constituted as is the one that Descartes envisioned, a mechanism inhabited by consciousness.[15] Consciousness is a secondary consideration in as much as it is not on the 'scale of the gaze.' Consciousness is what reveals, initially, the disfunctioning of the body through complaints of pain, shortness of breath, etc., but it is the body to which medicine addresses itself, to which the discursive gaze is oriented. The body, unlike consciousness, is a device that can be manipulated, worked on, *fixed*. (Patients with symptoms that cannot be correlated with physical findings or categorizable disease processes, i.e., symptoms which are not on the 'scale of the gaze,' are labeled as 'psychosomatics' and shipped to the psychiatrist.) What is invading the device can be killed, malfunctioning parts that are inessential can be cut out (e.g., gall bladder, spleen) malfunctioning parts that are essential can either be replaced (transplantation) or have their operations taken over by drugs (e.g., insulin in the case of the pancreas) or by machines (e.g., dialysis in the case of the kidneys).[16]

The last two types of treatment (transplantation and mechanical or pharmaceutical replacement of organ functions) are interesting because of their relation to disease and the life of the body outside of the medical regime, i.e., the other discourse/practices that the (now) patient inhabits. Such treatments do not, and indeed cannot, have as their goal the return of the body to its former life. That is because they do not cure disease. Instead they redirect the flow of the disease process. Kidney transplant patients do not walk out of the hospital free of disease. On the one hand they no longer go through a regimen of dialysis (that is done anywhere from twice a week to daily). On the other hand they do have an entire regimen of medications, of avoidance (due to their immunosuppressive drugs they must avoid all sources of infection, people with colds or other infectious diseases, crowds, etc., since even minor infections can be catastrophic for them), self surveillance of the body for signs of rejection or side effects of their medications, etc. Similarly, diabetics and dialysis patients arrange their lives around their disease process (a process which includes its treatment).[17]

We must remember that the alternative for such individuals is no alternative at all, at least for most. While death may illuminate the body for medicine it remains the last fall of darkness for the rest of us. And that is the problem for a postmodern critique of medical science. Medicine is remarkably successful in its own way. It does 'fix what ails us' quite often. For most of us the physician is the one who sets broken bones, cures minor infections, alleviates allergies, patches us up after auto accidents. So, why ask the question in the first place? It is not simply a question of knowledge for its own sake. The purpose of postmodern thought, or at least its Foucaultian variations, is to understand so as order to produce shifts in the present order. We critique medicine, or anything else, in order to produce effects, in order to transform the present. However, given the effectiveness of medicine it is not clear that it needs transforming, i.e., it is not clear that it is broken. And, as any mechanic who lives technological devices and life will tell you, you should not try to fix 'what ain't broke.'

But the past decade has given us reason to be suspicious of medicine, suspicions that should, perhaps, have been raised before. The appearance of a disease that hid successfully from a 'gaze' that seemed to demurely divert its eyes and then was unable to control what it was forced to see has made us want to look under the skirts of a *techne* that we had trusted, and indeed want to trust, to see just what is supporting its public facade. Medicine's reaction to this disease has forced us to re-evaluate both its internal functioning and its position among the other discourses that form the fragmented and multiplicitous planes of our existence.

But that is a move we are reluctant to make since those are the very skirts that we desperately cling to, and perhaps even hide behind. Medicine's reluctancy in regard to AIDS (the specific genealogy of which reminds us even more of medicine's inextricable relations with its *milieu*, particularly its political, economic and moral *milieu*) has not dissuaded us from placing our hopes (and economic and political energy) with it. We still expect medicine to save us from the plague. And perhaps with good reason(s).

What Foucault's archaeo-genealogy does is to give us a story about what the structures and boundaries are of our current 'gaze.' What it cannot do is to extricate itself from that 'gaze' in order to glimpse the promised land. Current efforts at transforming the present with regards to AIDS involve attacking the socio-politico-moral structures which produce the disease as something which occurs to 'others' and its victims as modern day lepers. But these efforts are not an attempt to undermine the structures of the medical gaze. They are efforts at freeing this gaze, at allowing it to see its victims in a light that makes it want to help them, at generating the economic force that allows this gaze to look more vigorously than it has before, that allows it to look at, and act for, more objects of that knowledge that we desperately want to be revealed.

What then would a postmodern doctor look like? She would be a physician who realized the discurso-genealogic structure of medicine. This means (at least) two things. First, she would realize that the present form of medical discourse/practice is one that arose within a heterogenous field in which the lines of descent are multiplicitous. This is to say that medical knowledge is ruled or constituted within this field rather than standing on its own in the path of an empirico-phenomenological vision. Second, and as a result of this, she would realize that there are other stories to tell, other, at least possible, ways of practicing medicine. This is to say that she would not take her own discourse, her own vision, as the only proper way of understanding the patient and the disease.

All well and good, but what are the practical consequences of this? We might want to say that this will allow her to be more open, more receptive to the voices of her patients, to the voices of her allies within the health professions, to the clamoring voices of populations with a high rate of infection of particular diseases, to the voices of families, to all that speak and to all that do not or can not speak. Yet, as we have seen above, medicine already does just that, albeit in a way that reduces those voices to documents that are organized by the medical voice, that categorizes according them to its own syntax and grammar, that listens only to translate without regard for what is truly different in the original tongue.

There is another possibility. Since our postmodern physician is the one that declines belief in the absolute status of modern medicine she is the one who can seek other discourses for disease, other discourses for treatment, seek what Thomas Kuhn would call a paradigm shift.[18] But if we take Foucault as our guide for understanding the relation between discourse and the functioning of science this possibility also seems untenable. We need to see that it is precisely the discursive structure of medicine, regardless of its genealogic beginnings and its lack of correspondence to the Truth, that allows it to function as it does, that allows it to identify and treat disease, that allows it to expand its abilities. AIDS became a disease that could be recognized, that could be diagnosed, that could be treated (however limited that treatment is at the present), and whose transmission could be interrupted, precisely at the point when medicine constructed a discursive strategy that wove together the stories of rare diseases, immunologic suppression, transmissions of bodily fluids, to create a certain discursive object. And it is the insertion of the individual into this discourse that allows h/er to be(come) a doctor. And it is the leaving of this discourse that constitutes the leaving of medicine.

This last possibility makes the modernist mistake that many postmodernists repeat of assuming that since discourse is multiple and not tied to an underlying reality that we then can control the production and generation of discourse. But it is precisely Foucault's point that this is not possible. It is discourse that opens up possibilities for us, not the other way around. That is to say that it is in its discourse that medicine finds *both* its limitations and its efficacy. One of the problems that those afflicted with environmental illness (EI) have encountered is a lack of credulity from physicians with respect to their disease. Medicine in general, and many physicians in particular, find the narratives of this disease to be anecdotal and uncorrelatable with physical findings (i.e., no inscription devices can transform these diseases into written documents). While many of these people have found some treatment and relief from alternative sources what many more of them would like is to have medicine recognize their disease and turn the power and resources of its gaze to it in an attempt to cure it, treat it or provide the scientific basis for restricting some of its causes. But it is precisely the point that medicine can only address this disease insofar as it finds a strategy for translating it into the discourse of medicine, of reducing it to the parameters of its own knowledge.

Up to this point I have been addressing this question primarily with regard to the private relation between patient and physician. If I am right in what I have argued above then there is little difference in the way that a postmodern and a modern physician diagnose and treat disease precisely because both must constitute themselves within this discourse in order to *be* physicians. Further, if this is the case then there is no reason to fear the loss of the reality of our experience of our selves, our disease, our pain, in the discursive gaze of the postmodern physician any more than we should before the objectifying gaze of the modernist physician. Neither will, or can, retain what we experience as the reality of our lives and address us as patients, as bearers of disease, as knowable and treatable entities. Both necessarily exert a violence on the subject and the body in order to diagnose and render treatment.

But this is not the end of the question. The border between the private relation between patient and doctor and the functioning of medicine in the public sphere is a porous one that keeps very little inside it and excludes even less.

This is because medicine is more than a local knowledge, a discourse or meta-narrative restricted to hospitals and doctor's offices. Medicine has become a 'grand' meta-narrative, an idealized ideology that threatens to colonize all other discourses of life. As Foucault points out, as cited above, health has replaced salvation. (Foucault, 198) Our redemption is no longer organized around how we treat others or how we regard the divinity. It is organized around how we treat our bodies. The desire for eternal life in another realm has been replaced by the desire to indefinitely prolong and improve life in this realm. Our new priests are doctors, diet experts, exercise instructors; the cross has been replaced by the staff of Aesculapius.

Again, it is AIDS which brings us face to face with this medicalization of life. While medicine had for the last few decades prior to the epidemic increasingly placed sex more and more under the control of individuals (by disconnecting the choice of pregnancy from the choice of sex and by curing STDs) medicine is now reclaiming that power. If we fail to place the ways in which we show love and produce pleasure under the control of medical direction we are threatened with a lingering disease that tortures for a long time before it kills. It is here that we confront the possibility of our own pain and death and are forced constitute our selves in the quadrangle of power/knowledge/pleasure/death.

And it is here that our status as modern or postmodern patients confronts us. The modern approach is to submit to the discourse and organize our practices around it, to give ourselves over to the grand meta-narrative that promises us life, that promises us health, that promises to stave off death, if not forever then at least for a long while. But to do so is to submit to a new *scientia sexualis*, one no longer organized around contamination by the moral degeneration of perversity but around the viral contamination that can accompany any sexual activity that involves the transmission of fluids. But to make safety from contamination our priority is not simply to act in a different way. It is also to

166

speak in a different way and to think in a different way. As both Stephen Schecter and Susan Sontag show, allowing medicine to construct the borders around our sexuality is to reconstitute our lovers, both past, present, and future, as a danger, a threat and to reconstitute ourselves as a danger and threat to them.[19] It is to introduce alienation into the heart of that which once held the promise being the haven from all the estrangements which constitute modern and postmodern life. Medicine's gaze has become our gaze in which those to whom we are attracted are also those whom we see as dangerous and in which our own desire becomes the medium of our finitude.

It is here, then, that we can find a difference in the postmodern physician that makes a difference. It is the postmodern physician which can realize the local character of her discourse and its tendency to dominate other discourses, its tendency to provide the myths and metaphors that structure the public understanding of disease. It is such a doctor that can realize that referring to certain communities of people as 'high risk' groups (rather than as 'high victimization' or 'highly vulnerable' groups) constitutes these communities as threats, as enemies, as Other, as groups to be feared rather than as groups to be helped.[20] It is such a person that can introduce resistances, transgressions, disruptions, into the functioning of medical discourse in those places where that discourse wreaks a violence on bodies that is over and above the violence that is a necessary part of its functioning, in those places where this local narrative escapes its bounds and colonizes the rest of the discursive world. It is also such a doctor that resists the entry of this form of medical discourse into her office and her hospital.

So, one *does* want a postmodern doctor. But there is a problem with this answer. If one wants a postmodern doctor this is because it is this sort of doctor who avoids (or at least attempts to avoid) the illegitimate extension of the constituting effects of that discourse beyond its local borders and even attempts to disrupt the effects of that extension by the discourse itself (i.e., by the dissemination of effective statements by her modernist colleagues into other discursive formations). However, to seek out such a physician for this purpose is problematic in that it continues to leave the question of our identities in a modernist position. The modern approach, from the side of the patient, is to submit to the discourse of medicine, to organize our discourse/practice around it, to give ourselves over to the meta-narrative (grand or not) that promises us life, health and the indefinite postponement of death. And to ask the postmodern physician to save us from this discourse (and keep us healthy and alive) is to remain a modernist patient.

And it is here that the correlative question to whether one wants a postmodern doctor is raised, *viz.*, does one want to be a postmodern patient? Further, how is one to be a postmodern patient (and stay alive)? To be a postmodern patient is to refuse to accept medicine as a grand meta-narrative, to recognize its character as a local knowledge. But is it possible, e.g., to disengage the techniques of safe sex from the discourse and gaze of safe sex? The question of inhabiting, and being inhabited by, multiple discourses, of establishing rapid transit networks between them, raises the problem of being in the right discourse, at the right time, in the right way, to avoid death. After all, if we have a discourse that allows us to continue discoursing we should use it to our best advantage. But at the same time we must avoid falling into the trap of instrumental reason, in this case a grand narrative that directs us to what discourse we ought to be in now (when in doubt, go with medicine) and the illusion that we are in control of discourse, that there is a subject, prior to the identities constituted within discourse, that we can turn to in order to decide which discourse we ought to be in at any particular moment. The further question is whether or not we can avoid living according to a grand narrative, a discourse of discourses.

There is a sense that we must. To do otherwise is to hand ourselves and our identities over to medicine, to allow a tool that makes life easier to become that which defines life. This is a reduction that far exceeds the local reduction that occurs in the hospital, the doctor's office, the chart. Yet we must also beware of the encroachment of medicine's 'public' face in these 'private' areas. The signifiers AIDS, HIV+, IVDA (intravenous drug abuser), ETOH (alcohol abuser), homosexual, when placed on charts or nursing care plans serve not only to identify diagnoses or possible vectors of infection but also to constitute the identity of the patient as a threat, as Other, as someone whose disease is "their own fault".

In a discussion with Foucault, Gilles Deleuze likens a theory to a "box of tools".[21] By this he means that theories are not abstract entities that determine reality but tools for working on that reality. Medicine simply is a tool. It is a theoretical and material means of the production of health. It is a local knowledge that we can avail ourselves of in order to attain (when possible) certain ends. To that extent it is a local knowledge. But it is a local knowledge that long ago escaped its borders. When we enter this locality we hand ourselves over to it, at least in part. And this necessarily so. Given the current functioning of medicine we must enter the discourse, allow ourselves and our bodies to be defined by the discourses and practices, at least temporarily, partially, in order to extract from it the desired results. But we cannot allow medicine to define those results for us. To do so is to be captured by a single discourse, by a single rationality, by a single grand narrative, to allow it to dictate what life is rather than to demand of it to allow us to live other lives, to inhabit other discourses.

Does one want to be a postmodern patient? Well, I believe that the answer is 'yes', but unfortunately it is not as simple as that. In his last writings Foucault suggests that the response to the disciplinary structure of contemporary society is an 'aesthetics of existence,' a making of one's self a work of art. If we understand the self, the subject, to be an effect of discourses and practices then the means of doing this is to generate the perverse intersection, interpenetration, intermingling, commingling of the polymorphous discourse/practices that structure the various aspects of our lives in the hope of producing effects within our own lives and those of others. What we must beware of here is thinking of *this* discourse/practice as a simple procedure. Art, even postmodern art, is not something that is produced easily.

And being a postmodern patient is not easy either. It depends on the taking up of medical discourse in such a way that it can be intersected, interpenetrated, intermingled to produce a work of art, a work that is not simply the effect of a single discourse. But that means knowing how to move within the discourse, recognizing that we can understand the discourse, not turning away, or simply nodding our heads in the face of that discourse but taking it up, using it, and disrupting it when it does not apply, is wrong, illegitimately extends itself beyond its bounds. Becoming this peculiar sort of artwork, an artwork that is aware of itself (at least partially), that is the combination of artist and artwork, attempting to manipulate and articulate those discourses that manipulate and articulate us, is a Sisyphian task that can all too easily leave us under the rock and requires great effort to simply continue the effort.

The irony of this is that our lives, and our identities, *do* hang in the balance.

Notes

1. Foucault, M. (1975), *The Birth of the Clinic: An Archaeology of Medical Perception*, Sheridan-Smith, A.(trans.), Vintage Books, New York, p. 166. Herafter cited in text as Foucault, followed by the page number.

2. See Lyotard, J.-F. (1984), *The Postmodern Condition: A Report on Knowledge*, Bennington, G. and Massumi, B. (trans.), University of Minnesota Press., Minneapolis.

3. Foucault, M. (1972), *The Archaeology of Knowledge*, Sheridan-Smith, A. (trans.), Pantheon, New York.

4. This is also true of *Madness and Civilization*. The only historical work of Foucault's that is not genealogical is *The Order of Things*.

5. This is even more apparent in the case of the HIV (human immunodeficiency virus) test and AIDS. While a positive test for the HIV is commonly considered as the definitive diagnosis for AIDS it is

not. This is the case for several reasons. First, the test does not test for the presence of the HIV itself. It tests for the presence of HIV antibodies, that is for the reaction of the body to the presence of the HIV. Moreover, it does not test directly for that antibody but for the existence of a certain state of blood chemistry that is indicative of the presence of that antibody (a state which is also consistent with the presence of antibodies to malaria). Second, the presence of the HIV is not equivalent to the presence of AIDS. AIDS is diagnosed when the patient acquires the opportunistic diseases (*pneumocystis carinii pneumonia*, Karposi's sarcoma) that are characteristic of the syndrome. Lastly, the AIDS diagnosis still requires the identification of a history that reflects the possibility of infection by HIV (blood transfusions, certain sexual practices, etc.). AIDS is *an sich* a set of symptoms rather than an entity over and above those symptoms. See Patton, C. (1990), *Inventing AIDS*, Routledge, New York, pp. 32-4.

6.   See Latour, B. and Woolgar, S. (1986), *Laboratory Life: The Construction of Scientific Facts,* Princeton University Press, Princeton, p. 51 *et passim*, for a discussion of "inscription devices."

7.   What we must beware of in this story is the tendency to read Foucault as, what one might call (somewhat crudely), an historico-linguistic idealist. The point is not that the body, life, disease are brought into existence by discourse. The point is that they are produced as objects of knowledge, and thus of intervention, by discourse. That is, there is no immediate relation between consciousness and its objects. This relation is always mediated by discourse and all the various effects that a discourse's genealogical beginnings produce. Since we can show that discourse has generated certain objects, knowledges, in the past that we now know to be 'false' we can make a reasonable wager, not the least to be on the safe side, that the same operations are at work presently.

8.   Scarry, E. (1985), *The Body in Pain: The Making and Unmaking of the World*, Oxford University Press, New York.

9.   There is further the problem of translation between discourses. I once witnessed an AIDS patient at the terminal stages of *pneumocystis pneumonia* asking to be intubated, i.e, to have a tube placed in his throat that would enter his trachea and be hooked to a ventilator that would take over the effort of breathing for him. To have this done to one's body is a painful, almost tortuous, experience (and one that signaled the end of all hope for a life of anything other than one hooked to this machine) and

this patient was well aware of this. While anyone familiar with these devices and the practices that surround them will recognize the pain that that patient must have experienced in order make this request, it is difficult, if not impossible, to communicate this experience to someone outside of this discourse/practice.

10.   Scarry, pp. 6-8.

11.   Foucault, M. (1979), *Discipline and Punish: The Birth of the Prison*, Sheridan A. (trans.), Vintage Books, New York; (1980), *The History of Sexuality, Vol. I: An Introduction*, Hurley, R. (trans.), Vintage Books, New York.

12.   Foucault (1979), p. 29.

13.   Such documentation is, or has been, done under the anacronym of SOAP (Subjective, Objective, Assessment, Plan). Despite the obvious temptation I shall refrain from a discuusion of purifying the body of disease in the neutral space of the clinic with a SOAP note.

14.   The discursive constitution of the patient is already moving in the direction of the hyperreal simulation of the patient.

15.   In this respect see Descartes, "Meditation VI," *Meditations on First Philosophy* and *The Passions of the Soul* in *The Philosophical Works of Descartes, Vol. I*, Haldane, E.S. and Ross, G.R.T. (trans.), (1975), Cambridge University Press, New York, pp. 185-199 and 329-427 respectively.

16.   The treatment of the body as a machine is perhaps seen most clearly during cardio-pulmonary resuscitation. In this procedure the heart and lungs are forced to continue to work after they have failed by mechanical means (external massage [a euphemism for violent rhythmic pressure on the breastbone] in the case of the heart and the forcing of air into the lungs). This is accompanied (in clinical settings) by the administration of chemicals and, sometimes, electrical shock to correct abnormalities and stimulate the organs to resume their functioning.

17.   This restructuring of their life is done by teaching them discourse/practices which then must dominate their life in order to continue it.

18. Kuhn, T.S. (1970), *The Structure of Scientific Revolutions*, University of Chicago Press, Chicago.

19. See Schecter, S. (1990), *The AIDS Notebooks,* State University of New York Press, Albany , and Sontag, S. (1990), *Illness as Metaphor* and *AIDS and its Metaphors,* Anchor Books, New York.

20. See Stephen Watney's (1987) discussion of this in "The Subject of AIDS", *Copyright* 1, pp. 125-32.

21. See "Intellectuals and Power: A conversation between Michel Foucault and Gilles Deleuze," in *Language, Counter-Memory, Practice*, Bouchard, D.F. (ed.), (1977), Cornell University Press, Ithaca, 1977, p. 208.

10 Scientific Discipline and the Origins of Race: A Foucaultian Reading of the History of Biology

Ladelle McWhorter

Introduction

Galileo, bound, is led to a chamber where he is shown the machinery of torture his inquisitors have at their disposal – wheels, racks, screws, caldrons boiling with acrid substances, branding irons, pincers, chains. In similar rooms across Europe, tens of thousands of men, women, and children have already died after hours or even days of unimaginable pain. The inquisitors are all but omnipotent, and Galileo, seventy years old, is virtually alone. At the sight of naked power spread before him, his will to truth is broken. Kneeling, he repudiates his life's work.

Truth, we are told, triumphs. Science emerges victorious over superstition. Power is checked by reason. And now, from the vantage point of modernity, we may look back and condemn the inquisitorial oppressors, regret Galileo's weakness, and solemnly reaffirm the moral of the tale – truth and power are mortal enemies, and vigilance is necessary if truth is to prevail.

But is that so? Michel Foucault, for one, suggests that it is not. His work directly contradicts the old and venerated idea that power and knowledge stand fundamentally opposed. Power and knowledge are mutually supporting, he claims, and must be analyzed in the complexity of their interrelations. To that end, he introduces his analytic notion "power-knowledge."

Foucault's "power-knowledge" is a controversial concept. Brought into English-speaking theoretical circles less than two decades ago, its meaning and range of applicability are still in dispute. While no one denies that some fields of social scientific knowledge (such as criminology) intersect institutionally with mechanisms of power, these intersections do not seem, to many, to constitute any essential relation of "mutual reinforcement" between knowledge and power. If, in rare cases, politics and scientific research are admitted to be mutually constitutive, the results of their mingling are typically dismissed as propaganda or pseudo-science. A few thinkers are willing to allow the entirety of the human or social sciences to be dismissed in this way–thus leaving intact and untainted *science*, real science.

In the remainder of this essay, I will argue that at least one indisputably real science, biology, is analyzable as a series of structures of power-knowledge. I will contend that the science of life was both required and enabled by networks of power operating at the end of the eighteenth century. Further, once established as a distinct and reputable discipline, biology went on to create its own objects of knowledge whose management its researches were designed to perfect. One of those objects was race; much of nineteenth century biological research was aimed at categorizing what we would now call human phenotypes for the expressed purpose of managing and manipulating the current and future populations of the globe. Though some of this work can be dismissed as pseudo-science, much of it cannot. It was, quite simply, science. And it was, quite simply, completely entangled in the production of racist social structures. Biological science invented the concept of race as we know it today and so made possible the development of racial hierarchies in Western societies. Thus, natural scientific knowledge and power are not mortal enemies; they are partners.

Relations between power and knowledge: discipline

According to the modern account, Galileo was temporarily silenced by the Inquisition. His books were banned. His freedom of movement was restricted. Power acted as a limit on thought and deed, an agency of repression, a prohibition, a no. Many people believe that that is all power ever is, that power is always negative and never positively productive. But, in order to maintain that conception of power, one must make some very dubious distinctions. For example, one must draw a distinction between the forces of creativity and repression and reserve the name "power" for the latter alone, despite their obvious similarities. Further, one must insist that scientific activity is simply systematic discovery, not any kind of force at all. Since science discovers the truths that power would repress, on this view, science and power are oppositional; since science's objects are the creations of natural forces, science is not creative in itself. Thus one must posit a distinction between science, on the one hand, and art and humanistic pursuits, on the other, and also between science and technological creation.

Foucault does not leave this neat set of categories, oppositions, and distinctions intact; he exposes it as arbitrary. There are repressive forces and creative forces, but both are forces, powers, and they are interlocked; production represses, and repression produces. There is no pure form of discovery distinct from the power to repress or to create. Thus Foucault rejects this common, but dubious, account of power and proposes a different, more inclusive characterization of it. Power is not to be understood as a commodity the possession of which enables one person to repress another. Power is something that happens between people; it exists only in its exercise. Rather than some*thing*, power is better understood as a multiplicity of relations that constitute their own organizations; as the support these relations give each other, which enables the formation of networks or systems; as strategies "whose general design or institutional crystallization is embodied in the state apparatus, in the formulation of law, in the various local hegemonies,"(Foucault, 1978, 92-93) but whose origins do not lie there. In other words, power is relation, events of relating, whose repetition generates organizational networks of force-events, within which relatively stable institutions, objects, personalities, etc., may sometimes form. Thus, on Foucault's view, power relations are productive as well as repressive. They are productive of institutions, laws and prohibitions, but they are also productive of theories, ideas, practices, methods, beliefs, ways of behaving, ways of being who we are. Power is event, and within networks of repeating events, truths are formed.

If we accept Foucault's conception of power, the next question is exactly how and to what extent does scientific practice interact with it. We must be slow to make generalizations, but we can get some sense of ways in which power and knowledge might interact by looking at regional studies of disciplines. One such study is Foucault's account of criminology in *Discipline and Punish*, where he describes the creation of a human type, the delinquent, within the interplay of knowledge and power. This study is particularly instructive for my purpose here, for, as we will see, the discipline of biology also produced various human types – races, categories generated, like the category of delinquency, from sets of developmental norms. In preparation for my analysis of the creation of race and racial theory in biological science, therefore, I will offer an overview of Foucault's account of delinquency's creation in and through the science of criminology.

Criminology, a field of knowledge, intersects with judicial and penal exercise of power. Contrary to those who hold that knowledge and power are enemies, this intersection is not mere coexistence. Rather, Foucault claims, knowledge and power are "entangled."[1] He goes so far as to assert:

[that] power produces knowledge (and not simply by encouraging it because it serves power or by applying it because it is useful); that power and

knowledge directly imply one another; that there is no power relation without the correlative constitution of a field of knowledge, nor any knowledge that does not presuppose and constitute at the same time power relations. (Foucault, 1977, 27)

It is not just that criminologists and penal authorities happen to focus attention on the same people. Criminologists and penal authorities make each others' jobs possible. Knowledge extends the domains within which power-events can repeat themselves and produce effects; correlatively, power creates new objects of knowledge. This occurs in the carceral field through the dual processes of surveillance and normalization.

At the end of the eighteenth century, when imprisonment became the primary mode of punishment, surveillance as a technique of control was carried on within prison walls. At first its official purpose was only to help maintain order while the offenders did their penance, since the point of penitentiaries was, ostensibly at least, to give offenders occasion to reflect on their crimes, reflection supposedly leading to reform. But surveillance quickly gave rise to more interventionist mechanisms of reform. It made possible the rewarding or punishing of individuals for acts committed in prison, and, combined with the prison dossier, it made possible the modulation of sentences based on "good behavior" or its opposite. Thus, imprisonment and surveillance produced data, which could be turned into knowledge, which in turn enabled the extension of techniques of control. In this domain, knowledge and power are interdependent in that their conditions of extension include one another.

Foucault goes on, however, to make the more radical claim that not only are knowledge and power here reciprocally conditioned, but they are mutually constitutive. (Foucault, 1977, 183) To see his reasoning, we turn to his analysis of normalization.

The role norms play in structured processes of reformation is much the same as the role they play in medicine and pedagogy. By studying the dossiers of prisoners passing through a reformatory regime, scientists can generate norms and then classify prisoners according to their deviations from those norms. This is what Foucault calls "normalization."

Normalization does two things: It homogenizes, and it individualizes. Norms homogenize by enabling all difference to be understood and treated as deviation from the norm and therefore as essentially related to it. Everyone caught up in a normalizing system stands in some relation to the norm, for there is no outside to the classificatory system. At the same time, norms individualize, because they enable a precise characterization of each person who is to be classified. Exactly who a person is can be known by the intersection of all his or her deviations from relevant norms. One's individuality, then, is just the full set of one's deviations.

Once normalizing power produces the classification systems needed to isolate something like a "dangerous individual" (Foucault, 1977, 252), the intersection of a particular set of deviations, the delinquent is born. The delinquent is a type of self who exists "before the crime and even outside it" (Ibid.) much as a homosexual can be said to exist in the absence of any homosexual experience.[2] At a given time, the delinquent may already be guilty of crime, or he or she may simply have a latent predisposition toward it. Whichever, what the criminal justice system must address is not only or even primarily the acts the delinquent commits but rather this individual's tendency to commit illegal acts; what must be addressed is the individual him or herself. All kinds of intrusions into people's lives then justified on the basis of the fact that a family member, say, is a delinquent personality "at risk" for delinquent behaviors (such as drug abuse). Family members and others surrounding the latent delinquent can be disciplined to respond to him or her and to each other in prescribed ways to "manage" the risk of overt manifestation of delinquency. In the process, lives are shaped; simultaneously, more data are gathered and concepts and techniques for identification and management are refined. The interdependent extension of power and knowledge continues, justified now by the figure of the delinquent, a category constituted within a regime of power and its correlative knowledge system.

Delinquency, as an object of knowledge, is produced, Foucault claims, by the carceral system. It is an epistemic object that is the product of the exercise of a certain form of power. Thus, power is in fact constitutive of knowledge, at least where criminology is concerned. Further, since penal institutions (along with the carceral system more generally) exercise power in relation to delinquent individuals, knowledge of such individuals is essential for the system's functioning. Power could not be exercised – which means it could not exist–in the absence of such knowledge. Therefore, just as power is constitutive of knowledge, knowledge is constitutive of power. In the discipline of criminology, Foucault shows us, knowledge and power are completely interdependent. But is the same true in *real* science?

The rest of this essay is concerned with showing how the concept power-knowledge can be applied to analysis of biology. I will draw on an early work, *The Order of Things*, to characterize biological science in its incipience. I will argue that the scientific debate over the origin and classification of variant human morphologies actually produced the epistemic object race and was instrumental in establishing nineteenth century racisms. This process of epistemic production parallels the production of delinquency in significant ways and serves similar purposes, such as the management and discipline of populations.

The origins of biological science

Biology is the study of life. Its predecessor, natural history, is the study of the orders of natural beings. It is a mistake, Foucault contends, to see the former as a simple outgrowth of the latter. He writes,

> Historians want to write histories of biology in the eighteenth century; but they do not realize that biology did not exist then, and that the pattern of knowledge that has been familiar to us for a hundred and fifty years is not valid for a previous period. And that, if biology was unknown, there was a very simple reason for it: life itself did not exist. All that existed was living beings, which were viewed through a grid of knowledge constituted by *natural history*. (Foucault, 1970, 127-8)

Foucault sees an epistemic break between natural history and biology, a break that can only be understood upon examination of the broader epistemic grids that existed in the eighteenth and nineteenth centuries. Natural history is to be understood against the background of the "classical episteme," wherein knowledge consists of tables of identities and differences. This way of understanding the world precludes any concept of "life," because a tabular notion of reality insists that nature is perfectly continuous, whereas "life" names a qualitative leap in the order of things.[3]

Foucault locates this major epistemic break just at the end of the eighteenth century. If that is plausible, it is no surprise that the word "biology" was not introduced into scientific discourse until 1802.[4] Biology, the science of life, was not possible until the category "life" had been formulated, a formulation inextricably bound up with notions of temporality, mortality, limit, sequential change. These are the ordering principles of a new episteme, one that does not view the natural world as a static, continuous tabular plane. Biology is essentially the study of processes and functional norms. It is a science wherein human being is primarily

> a being possessing *functions* – receiving stimuli ... reacting to them, adapting himself, evolving, submitting to the demands of an environment, coming to terms with the modifications it imposes, seeking to erase imbalances, acting in accordance with regularities, having, in short, conditions of existence and the possibility of finding average *norms* of adjustment which permit him to perform his functions. (Foucault, 1970, 357, *Foucault's italics*)

Although Foucault wrote the above words years before he began studying

178

normalizing power, biology appears here in its similarity with psychology, criminology, and other disciplines whose aim is to discover the norms of function and development of a given type of being within a given context. What remains is to examine how specific developments within biological knowledge are related to developments in power.

Biology and bio-power

Concurrent with the rise of biology, at the turn of the nineteenth century, Foucault contends, disciplinary power was developing its techniques of normalization. Normalizing disciplinary power is a positive force; it is the power to posit, to shape, to cultivate, to create. Any limits it imposes are in the service of production and growth. Its primary concern is the generation, maintenance, and management of living human subjects. It is preeminently an administrative power.

> The old power of death that symbolized sovereign power was now carefully supplanted by the administration of bodies and the calculated management of life. During the classical period, there was a rapid development of various disciplines – universities, secondary schools, barracks, workshops; there was also the emergence, in the field of political practices and economic observation, of the problems of birthrate, longevity, public health, housing, and migration. Hence there was an explosion of numerous and diverse techniques for achieving the subjugation of bodies and the control of populations, marking the beginning of an era of "bio-power." (Foucault, 1978, 139-40)

Effective management strategies presuppose knowledge of the objects to be managed. If the goal is to manage birthrates, longevity, health, and migration – processes – it is important to understand the physiological processes of human beings. Thus, by the beginning of the nineteenth century, a science of such processes, a science of something like life, was imperative; biological knowledge was absolutely essential for the effective exercise of disciplinary power.

Along with the emergence of this administrative imperative to manage the physiological functions of populations, the concept of life as physical, material organic finitude arose as a viable epistemic category. A science of life not only became necessary but also became for the first time possible, since its condition of possibility, the object "life," had been marked out within a field of power relations. Hence, biology is not only a necessary tool for disciplinary power; the very possibility of a science of biology is actually constituted by that power.

179

As a normalizing discipline, then, biology is an arm of bio-power. Once constituted and legitimated, biology aided (and still aids) the extension of bio-power in numerous ways. In the last century, one prominent way was the production of the population management tool, race.

Before the nineteenth century, the word "race," in both English and French, meant "lineage." To be a member of a race was to have a certain heritage; it was a matter of descent, though it was not yet linked to anything like a "genetic heritage."[5] One could speak of the Greek race, the Jewish race, the Russian race, or even races of animals such as the Dalmatian race of dogs. Though race might incidentally correlate with morphology, its primary sense was that of breeding; it had some of the same connotations that the word "breeding" has in common English usage – designating both one's parentage and one's manners.

The term "race" was not reserved for human phenotypic variety until the nineteenth century. It is only during that period that race and racism as we know them today came to be. That fact is not obvious now mainly because we are used to understanding slavery as a product of racism, which means racism must have existed prior to the nineteenth century. However, as I will show, racism is an effect, not a cause, of slavery.

Prior to the nineteenth century, during the European expansion into Africa and the Americas, justification for conquest and slavery did not come from racial theory but rather from theology; the operative distinction between Europeans and others was religious: "we" are Christian; "they" are pagan.[6] By the late eighteenth century, though, the large and growing number of slaves who converted to Christianity posed a problem for their owners; the requisite emancipation would constitute a serious financial loss and would produce a free and perhaps unmanageable underclass in colonial territories. A new, non-theological justification for slavery was needed.

Until that time slavery was not linked to skin color or body form as race is. What we now call "white slavery" was common in the fourteenth and fifteenth centuries when many Slavs were sold in markets in Italy, Spain, Egypt, and the Mediterranean islands. It was not until 1793, when the expansion of Russia, culminating in the annexation of the Crimea, cut off the supply of white slaves to Islamic markets, that "black" could become synonymous with "slave."[7] Only then could morphology possibly be taken as the basis for slavery's justification. That possibility was realized through the creation of sciences of race within the new discipline of biology.

In 1799, Louis Francois Jauffret founded the Societé des Observateurs de l'Homme. (Stocking, 15). Its members were to observe various groups of human beings and study the gradations in those groups from primitivity to modernity. (Stocking, 27) The Societé called for classification of human groups based on comparative anatomy, comparative study of customs, construction of a typology of France to determine the influence of climate on body form and habit, founding of a museum of comparative ethnography, and compilation of a comparative dictionary of all known languages. (Stocking, 16) The discipline we now call anthropology thus emerged as a subfield within the life sciences. It had two goals: to establish a classificatory system and to establish a hierarchy of development among the world's human groups.

Societé members sailed to Africa, Tasmania, and New Holland (now Australia) in early 1800 in order to observe those regions' aboriginal peoples.[8] Others continued the work that the Societé began; later observers concentrated on various types of somatometry, with head-form becoming the main morphological trait studied, in part because of the possibility of comparisons between the skulls of living and pre-historic people. At first the study of head-form was undertaken in order to prove either polygeny (the view that humanity is a set of species with separate origins) or its opposite, monogeny.[9] But as the century progressed, the question of whether *Homo sapiens* was one or several species gave way to the questions of how many races there were and how best to characterize them. Race became the preeminent object of bio-anthropological study.

Before we examine how the object race came into being, it is important to emphasize that Jauffret and his colleagues and successors were natural, not social scientists. What we would now call physical anthropology was part of biology, and its practitioners were anatomists, physiologists, zoologists, and medical doctors. Though anthropology became a separate department in many American universities between 1890 and 1910, until it ceased attributing most significant human variations to hereditary factors, it could not function as a discipline theoretically and methodologically separate from biology. It did not develop genuinely distinct theories and methods until the 1920s. During the period under consideration here, the theories and methods of anthropology were completely consistent with the theories and methods of other branches of biological science.

Exactly how type or race became an object of study at the same time that scientists were emphasizing temporal organic processes is difficult at first to understand; on the surface, the predominance of typology seems in direct conflict with biology's emphasis on function and process; the establishment of static identities seems in direct conflict with biology's emphasis on change through time. The key to understanding lies in the fact that nineteenth century scientists, from the very beginning of biological study, were interested in the history of human development. Developmental thinking, far from precluding typologies, made them all the more important, because development was seen to occur in stages. For example, the fact that a human being develops from a cluster of cells quite unlike any living person led biologists to see bodies as the culmination of a process that they might understand through study of the stages of fetal development. While stages are of course merely transient formations in process, biologists tended to conceive of them as discrete – and *static* – moments in a developmental chain.

Thus, stages were easily construed as types; and, conversely, types, the morphological variations biologists observed, must be stages. If those stages could be isolated and studied, then perhaps the full process could be delineated. While this seems harmless enough in relation to comparative anatomical studies of dead fetuses, we must remember that the larger project was that of describing the development of human civilization. Adult "types" – races in particular – were believed to be stages – arrested and permanent, perhaps, but *stages* – that could serve as clues to the mystery of human development.[10] Thus was biology a major force in the creation of the concept of race as graded type. Superior and inferior human types – races – became facts. Biology then set out to take account of those facts. Data on morphological groups were amassed and norms of development were established. Races were ranked according to how civilized (or uncivilized) their representatives were thought to be.

By 1850, the anatomist Robert Knox could argue successfully that races were the result of arrested or deviant development; race occurred when there was a retardation of normal developmental processes. (Knox thought Saxons were the only people who were not retarded; all other groups exhibited some form of developmental deviation.)[11] By Knox's time developmental thinking was deeply rooted in natural science. Typology was no longer a matter of classification as it had been in natural history. Type was a function of normalization, and the variations characteristic of types were deviations from norms. Though many people disagreed with Knox's particular racial hierarchy, his framework for understanding race – that it had to do with normed development – was perfectly consistent with the science of his time.[12]

Once race (meaning those morphologies that differ from the Saxons, or whichever group puts itself at the top) came to be understood as deviation, the old concept of degeneration found its way back into natural science. In times past, degeneration was connected to the metaphysical notion of a Great Chain of Being emanating from a divine source; beings further from the source were more degenerate than those closer. But degeneration's theological origins became irrelevant as the concept's medical and social utility became apparent for the management of populations.

A degenerate individual was defined as one who failed to advance appropriately along normal physical, mental, and moral developmental lines. Since degeneracy was believed heritable and progressive, steps had to be taken to control those individuals and groups who exhibited any symptoms of it. Many people judged degenerate were prohibited by law from marrying; those judged likely to propagate outside of marriage (namely, the criminal, feeble-

182

minded, and insane) were sterilized. Race itself was a mark of degeneracy, so entire races were judged more or less degenerate based on the marks of race alone. However, scientists usually found plenty of other marks of degeneracy in those whose race indicated its presence as well. For example, the U.S. medical establishment "confirmed," using faulty data from the eighth, ninth, and tenth census, that the Negro race was dying out due to "physical degeneracy."(Gilman, 39) It was only a matter of time before that race destroyed itself. Other signs of degeneracy, insanity and perversion, supposedly increased by 1000% in African-Americans between 1860 and 1890. Some of this increase was thought to be due to the sexual freedom that emancipation supposedly brought, but unrestrained sexuality, scientists believed, would only cause further degeneration of the race. The only outcomes possible were mass sterility or mass insanity, (Gilman, 39) the preludes to natural racial annihilation.

Since degeneracy is just another name for deviation from developmental norms, and since criminality is one type of deviation from those norms, degeneracy supposedly included and produced an increase in criminality. (At this point biology and criminology were interlocked.) While awaiting the natural final solution, white Americans and Europeans had to be vigilant lest the criminal tendencies of inferior races lead to the injury, and the corruption, of their own. Management of racially-marked criminal populations was necessary; fortunately, criminals bearing the marks of race were relatively easily managed, since surveillance was relatively easy.

Because degenerates (of all sorts) were deemed to be doomed anyway, many scientists believed it was morally permissible to use them as experimental subjects, even if such use endangered their health or their very lives. Thus were members of allegedly degenerate races, as well as mental patients and homosexuals, used as test subjects. In one infamous case, scientists injected African-American subjects with syphilis and allowed them to become terminally ill; they considered this permissible because, they claimed, degenerate Negroes would have contracted syphilis anyway. (Gilman, 45)

With the publication of *Hereditary Genius* in 1869, Francis Galton inaugurated the eugenics movement, a bio-political movement that flourished in Europe and Japan as well as in America into the 1930s. Historians credit this movement with racist anti-immigration laws in the United States and Europe and with the forced sterilization of thousands of people. The purpose of the movement was to rid humanity of so-called defects, deviations from the norms of biological and social function established by scientific research. These norms – having been established by the same sciences that produced the epistemic object race – inevitably followed racial lines and, hence, were inevitably racist.

Thus, biology created the conditions for the possibility of racism. While many people dislike associating with those whose customs, religious beliefs, or

183

political views differ from their own, are disturbed by what is unfamiliar, and feel challenged in unpleasant ways by people whose behavior they find difficult to predict, there is no necessity that these feelings of discomfort, dislike, or fear should develop into racism. For that to happen, it was first necessary to invent the concept of race as a way of categorizing and naming difference. Since races were conceived within biology as stages along a developmental continuum, they were hierarchized from their inception. Thus, differential treatment of members of "inferior races," whether punitive or condescending, was a possibility created by scientific discourse, a possibility whose realization was inevitable.

Conclusions

Just as power-knowledge gave rise to the epistemic object delinquency and then used that object to extend the grasp of power, biological science (generously funded by governments intent on gaining socially useful knowledge for the efficient management of life) produced the epistemic object race and then, using it as justification, participated in the extension of power through the racial management of populations and individuals. The concept of race was very useful. It justified slavery. It justified hundreds of wars. It kept apart underclasses who might have joined together against the classes that ruled them. And, of course, it served for decades as an object of well-funded study for many scientists.

Of course, that was the past. Now, though racism is far from declining in the West, the biological concept of race is in disfavor. Genetic theory posits gene pools and populations as its objects of analysis, objects to which morphological races do not correspond; anthropologists now speak of ethnic groups. Therefore, some would argue, since scientific theory and racist practice are separable, science cannot have been responsible for racism. The flaws in this argument are too obvious to require enumeration, but it does point to an interesting phenomenon. In fact, scientific texts and theories do seem to be abandoning the concept of race and doing so at exactly the same time that racial minorities are using the concept as a rallying point for political action on their own behalf.[13] Perhaps the concept of race has outlived its usefulness as a population management tool and now poses more of a danger to the status quo than it obviates. At the end of *Discipline and Punish*, Foucault advances a similar speculation about the concept of delinquency; he suggests that the management of populations may no longer require the figure of the delinquent, who may be replaced with some other form of discipline or object of knowledge. (One might argue that in major American cities, delinquent populations, like racial populations, now pose a threat to the powers that posited them and so may have to be dismantled if those networks of power are to maintain themselves.) In any case, one should be very cautious about assuming that science is more "progressive" than society and very circumspect about the managerial intentions embedded in any new theory of human variation or development. Despite the stories illustrating the contrary, power and knowledge are intertwined. Even Galileo did not contemplate the heavens with apolitical objectivity; he worked in an arsenal in Venice.[14]

If my analysis is plausible, then Foucault's notion of power-knowledge is a useful analytic tool for application to at least one of the natural sciences. Those who would argue that sciences like biology are fields of knowledge only externally related to power are mistaken. Power permeates and shapes biology in ways very similar to the ways in which it permeates and shapes criminology, psychiatry, pedagogy, and industrial psychology. It created the conditions for biological science to come into existence when it created the category "life." Biology, like criminology, generates new epistemic objects, such as race, which it then claims the right to observe. Through extended observation it establishes norms of development and functioning; based on those norms it allows and sometimes even arranges for the management of deviation. Therefore biology, like the fields of knowledge Foucault studies in *Discipline and Punish*, is a normalizing discipline that arises within and extends bio-power.

Acknowledgements

I owe special thanks to Mark Bandas for his helpful suggestions regarding historical material, to Douglas MacPherson for his diligent bibliographic research and careful proof-reading, and to Debra Bergoffen for her extremely incisive editorial advice.

References

Banton, M. (1987), "The Classification of Races in Europe and North America: 1700-1850," *International Social Science Journal*, vol. 39.
Benedict, R. (1945), *Race: Science and Politics*, The Viking Press, New York.
Foucault, M. (1978), *The History of Sexuality, Volume One: An Introduction*, Hurley, R. (trans.), Vintage Books, New York.
Foucault, M. (1977), *Discipline and Punish: The Birth of the Prison*, Sheridan, A. (trans.), Vintage Books, New York.
Foucault, M. (1970), *The Order of Things*, Vintage Books, New York.
Gilman, S. (1983), "Degeneracy and Race in the Nineteenth Century: The Impact

of Clinical Medicine," *The Journal of Ethnic Studies*, vol. 10, no. 4.

Gould, S. J. (1981), *The Mismeasure of Man*, W.W. Norton, New York.

Harding, S. (1991), *Whose Science? Whose Knowledge?*, Cornell University Press, Ithaca.

Snyder, L. L. (1962), *The Idea of Racialism*, Van Nostrand Reinhold Co., New York.

Stepan, N. (1982), *The Idea of Race in Science,* Macmillan, London.

Stocking, G. W. (1968), *Race, Culture, and Evolution: Essays in the History of Anthropology*, Free Press, New York.

Notes

1.   Foucault (1977), p. 23.

2.   For a discussion of the homosexual as a type of self, see Foucault (1978), p. 43.

3.   The old Natural Historians believed all natural beings are locatable on one conceptual plane or table. Their tenet was "nature never leaps."

4.   Stepan, Nancy (1982), *The Idea of Race in Science,* Macmillan, London, p. 5.

5.   Banton, p. 51. Buffon had used the word "race" in a technical sense to refer to human variation as early as 1749, but his usage did not catch on. See Snyder, p. 11.

6.   Benedict, pp. 108-9.

7.   Davis, David Brion, "Slaves in Islam," *New York Review of Books*, Oct. 11, 1990. Davis goes on to say that racial slavery seems to have been a contribution of the Arabs to the Western world. We might also note that the English word "slave" derives from the name of a Slavic people enslaved in southeastern Europe.

8.   Stocking gives a fairly detailed account of the expedition. "Observation" actually involved by a great deal of interference, including sexual assault and rape of native women. Within thirty years after the Tasmanians were "observed" by European scientists, they became extinct. This lent credence to the longstanding view that Tasmanians were the lowest race on the ladder of civilization.

9.   Stepan, pp. 9-10. Joseph Camper seems to have started this, and Blumenbach tried to counter his work. Eventually craniometry was standardized with Retzius' development of the cephalic index in 1860. There were attempts to develop measures of skin color, but some, like A.L. Kroeber, objected that this could never be precise. The nasal index was also a popular measurement. See Snyder, pp. 14-17. Etienne Serres argued early in the nineteenth century that Africans are more primitive than Europeans because the distance their navels and penises (presumably only those of the males) remains shorter relative to body weight throughout life. See Gilman, p. 41, or Gould, p. 40. Craniometry was not fully discredited until Franz Boas' 1911 article showing that head shape varied with environment rather than being fixed by heredity; Boas also showed that stature varied with nutrition. Somatometry was largely replaced in the twentieth century by intelligence testing, but there are a few current attempts to revitalize it. Psychologist J.P. Rushton has argued recently that brain weight is an indicator of intelligence and that blacks and women of all races, having lighter brains than white males, are in fact less intelligent. See Rushton, "Race differences in behaviour: a review and evolutionary analysis," *Personality and Individual Differences*, 9:1035-40. For a discussion of Rushton, see Cernovsky, Z. Z., "Race and Brain Weight: A Note on J.P. Rushton's Conclusions," *Psychological Reports*, (1990), vol. 66, pp. 337-8, and Rushton's reply immediately following on pages 339-66. Cernovsky responds in "Intelligence and Race: Further Comments on J.P. Rushton's Work," *Psychological Reports*, (1991), vol. 68, pp. 481-2.

10.   The change in the meaning of the term "race" in both French and English was clearly enabled by and reflected in the work of natural historians and biologists. Cuvier conflates the notion of type and lineage as early as 1817 in *The Animal Kingdom*. There he groups together beings with distinct similarities, thus positing types, some of which are also groups of beings with similar lineages. The notion of type, then, *seems* consistent with the older notion of lineage, despite the fact that it is not. See Banton, 51-2. Cuvier, of course, was not interested in suggesting that types are stages on a developmental continuum (since he feared such thinking would lead to the heresy of evolutionism), but his rival Geoffroy, who believed that God has only one (or a very few) architectural designs which he varies to produce species, was interested in just such a possibility. See Appel, T. A. (1987) *The Cuvier-Geoffroy Debate: French Biology in the Decades Before Darwin*, Oxford University Press, New York, for an account of these thinkers' work. That

developmental thinking led to the formulation of type as arrested stage is evident in Geoffroy's science of teratology, the study of monsters. Just as species might be seen as various stages of development of one divine organic architectural design, deformed individuals might be seen as the unfortunate results of arrested or disrupted development within one species. William Ripley – usually considered now as nothing more than a circus side-show aficionado – was a serious teratologist and a student of race. Along with many of his contemporaries, he thought racial types – like congenitally deformed individuals – represented arrested stages of development on one hierarchical continuum.

11.   Stepan, pp. 41-44. I would like to note that this conception of race may account for the phenomenon often mentioned by those who charge white feminist and leftist theorists with racism. White theorists often treat non-white people as having a race, whereas they themselves are "race-less." If my claims about what race is are right, then in fact those white theorists *are* raceless, and their racelessness is one aspect of racism.

12.   The *OED* lists R. Knox as the author of an 1831 article in *Coquet's Anatomy* that uses the word "organic" in its modern bio-chemical sense. I have not been able to confirm that R. Knox and Robert Knox are the same man, but Robert Knox was at the height of his anatomical career in 1831, when the article came out. If Robert Knox is R. Knox, then there is a direct and interesting link in England between the solidification of the dividing line between organic and inorganic structure, on the one hand, and normalizing thinking, on the other. See Gillispie, C. C. (ed.), (1973), *Dictionary of Scientific Biography*, Vol. VII, Charles Scribner's Sons, New York, pp. 414-16.

13.   See Edlin, G., "Reducing Racial and Ethnic Prejudice by Presenting a Few Facts of Genetics," *The American Biology Teacher*, (Nov./Dec., 1990), vol. 52, no. 8, pp. 504-6, and Littlefield, A., Lieberman, L., and Reynolds, L.T., "Redefining Race: The Potential Demise of a Concept in Physical Anthropology," *Current Anthropology*, vol. 23, no. 6 (Dec., 1982), pp. 641-55.

14.   See Harding, p. 30.

11   Nature as Other: A Hermeneutical Approach to Science

Felix O'Murchadha

Introduction

The otherness of nature is beyond us, beyond our humanity, beyond history. Yet it is in a place carved out of nature that we dwell (*éthos*). In dwelling – as ethical beings – we become what we are. Hence, how our dwelling is conceived, how the relation of the history of this dwelling to the natural environment, through which it is carved out, occurs, relates directly to who we are. Natural science is the dominant relation to nature in the modern world. It follows that natural science is not merely a disinterested form of knowledge or on the other hand a form of knowledge geared solely towards utility, but rather helps constitute the truth, the true discourses, about us human beings, to the extent to which we are moderns. Nature is other certainly. It is not other as a person is other. This, however, only settles the question of its ethical relevance to the extent to which the decision has already been made as to the scope of ethics. The ethical is not the dwelling as such but the questioning dwelling. It is in encountering the other that the question of ourselves is brought to the fore. The question thus raised is not simply one regarding who we are as moderns, but rather how we become who we are. Hence, in approaching the natural sciences questions of truth and ethics are central, precisely because the natural sciences are one answer to the otherness of nature, which determines how we dwell and who we are.[1]

Hermeneutical philosophy, as exemplified by Gadamer and Ricoeur, by remaining committed to the *Natur-Geisteswissenschaften* distinction, is for the most part prejudiced against conceiving nature in this way. The other is generally conceived, in Charles Taylor's term, as a self-interpreting animal. This amounts to making an ontological distinction on the basis of the human ability to interpret herself, i.e. the ability to be self-constituting. I will argue that this distinction, which, in effect, expresses a bias in favour of a being who constitutes herself, a being not very far removed from the free subject of modernity, is undercut by the logic of the role of the other in the hermeneutical circle. In this paper I shall argue that the logic of certain hermeneutical concepts tied to a pragmatic concept of truth can help us reconceive the role of the philosophy of natural science in a postmodernist guise.

The role of the other is I shall argue a dialogical role and indeed it follows for my argument to work that every science has a basic if sedimented dialogical connection with its area of being. Every science opens up its own realm of being and this opening, the manner in which it is effected and the standards of truth which govern it, define the science. We say, in a very revealing phrase, that a certain area of being 'lends itself' to a certain form of investigation. This 'lending of itself' is the revelation which cuts out a place for a particular science. This place that is cut out is where the truth of being is revealed as physical or as chemical or as psychological etc. This is the context of the hermeneutical circle in each science. But the hermeneutical circle allows of degrees. The circle is constituted by openness, but this openness may take different forms. To the extent to which the interest in domination and control underlies the science, then the openness will be limited. It is, I think, for this reason that Gadamer makes a distinction between the natural and the human sciences on the basis of the interest in control (*Beherrschung*) in the latter, which he states the *Geisteswissenschaften* must guard against.[2] The openness basic to the hermeneutical circle is what Gadamer seeks to protect. For there to be a true hermeneutic of the natural sciences, this same openness needs to be sought there as well. To the extent to which Gadamer is right that the interest in mastery pervades the natural sciences, and to the extent that this interest forecloses the openness of which he speaks, then we are left with the problem of the status of such a hermeneutics of the natural sciences. Gadamer makes clear that he rejects a prescriptivist role for hermeneutics.[3] Yet, as we shall see, he appeals to the 'moral phenomenon' of the object of hermeneutical research, which indicates that in practice there is a type of prescriptivism in his approach. This prescriptivism involves an injunction to respect the moral significance of the object of study.

It seems clear that Gadamer sees the hermeneutical experience under threat to the extent that practical wisdom is eroded by the increased control of *techné*. Hermeneutics in his later writings has become a vehicle of critique. I shall argue that the logic of the hermeneutical position is that it becomes a critique of the natural sciences also on the basis of the same interest which motivates it in the human sciences: openness to the other.[4]

Science and human practice

The starting point of any hermeneutical discussion is the acknowledgement of the situation in which any being who wishes to understand finds herself. This situation is often termed simply the 'hermeneutical situation.' Essentially this situation is characterised by the fact that one always encounters that which calls for understanding with presuppositions. The situation of understanding is one of preunderstanding. Gadamer expresses this idea by stating that 'we are always underway' (*wir sind immer unterwegs*). Strictly speaking there is no starting point in the act of interpretation. Further, this absence of a starting point points to the everpresence of meaning. In other words, we never encounter a mere object which we then subsequently attempt to interpret and understand, rather what we encounter we encounter *as* something. This 'as structure' precedes explicit interpretation. We are never without preunderstanding because even that which we do not understand has meaning; it refers, that is, to the wider context of our understanding of the world.

The bringing together here of an image of movement (we are always underway) with 'meaning,' which might be conceived as stationary, is not accidental. It points to the fact that the 'hermeneutical situation' is not to be understood as a *theoretical* starting point from which the practice of interpretation can take its standards, rather it refers to the *practical* encounter with that which is worthy of question. Hermeneutics after Heidegger and Gadamer begins not with the specialised field of interpreting texts but with the practice of understanding. In other words, its concern is not with representing the world as it is, but, rather, with the rationality required in a situation in which one is faced with what one does not understand. Its concern then is with the practice of understanding.

The inescapability of this practice of understanding comes from the very nature of human practice itself. Meaning, as I mentioned above, is what characterises the situation. In other words, the situation of human practice is one in which all human pursuits are symbolically constituted, as such are interpretative, and are only accessible on the basis of how beings come to language. Through language human practice is accessible, because human practice is irreducibly meaningful. It is not that we first make the decision that we wish to understand, rather we could not exist humanly without understanding. The world addresses us. We find ourselves in the situation of being the addressee. It is to clarify this basic situation that Gadamer thematises experience.

In what is a pivotal section of *Truth and Method*, the analysis of historically effected consciousness (*wirkungsgeschichtliches Bewußtsein*), Gadamer discusses 'experience.' Although Gadamer uses Hegel's analysis of experience as negativity, his main concern is with experience in the sense of the experienced man of Aristotle's *Nicomachean Ethics*. I cannot here go into the full analysis of Gadamer's use of experience. For my purposes what is important is that the appeal is to a practical experience, in the sense that experience is always of the unexpected, of that which our past experience helps us to understand, but does not give us the basis from which to simply derive a solution to the problem – the lack of understanding – which our new experience faces us with. Gadamer characterises this experience as dialogical in the sense of involving a 'thou.' The 'object' of the new experience is characterised by not being totally objectifiable, but rather as being a Thou (*Du*). He defines a Thou as that which relates itself to us, to the interpreter. Gadamer is quick to clarify that no appeal is being made to the intention of the author in the manner of Romantic Hermeneutics. Nevertheless, basic to the interpretation of a particular text is the relation of dialogue with tradition. Tradition is the Thou, that with which we are in a constant relation of dialogue.[5] The implications of this with regard to the nature of hermeneutic experience (*Erfahrung*), are clearly spelled out by Gadamer as follows: "Since here the object of experience is a person, this kind of experience is a *moral phenomenon* – as is the knowledge acquired through experience, the understanding of the other person."[6] This entails that there are moral criteria to be applied to the approach suitable for the human sciences. In order to clarify what this approach should be Gadamer, I think significantly, appeals to Kant. He does, however, go beyond Kant by conceiving of this 'moral phenomenon' in terms of the openness to tradition characteristic of historically effected consciousness, which he claims to be the highest form of hermeneutical experience. This openness to the other Gadamer defines as the recognition that "I myself must accept some things that are against me, even though no one else forces me to do so."[7] Gadamer concludes the section on hermeneutical experience by stating:

> The hermeneutical consciousness culminates not in methodological sureness of itself, but in the same readiness for experience that distinguishes the experienced man from the man captivated by dogma. As we can now say more exactly in terms of the concept of experience, this readiness is what distinguishes historically effected consciousness.[8]

From this passage it is clear that the human sciences for Gadamer are deeply historical. Not only are they historical in the sense that they examine human beings, who are historical beings, more fundamentally hermeneutics is historical in the sense that it starts from an experience of tradition which is historically effected in the sense just described. There are no objects of research, but, rather, partners in dialogue, a dialogue which is always shifting, fuelled as it is by the dynamic logic of question and answer. The status of the texts as other and their historicity goes hand in hand for Gadamer. He can for that reason conceive the 'moral phenomenon' quite easily as a matter only of the human sciences. My interest here, however, is to draw out the implications of this moral nature of the scientific relationship in terms of the natural sciences.

One difficulty here of course is that while a dialogical model may be plausible in the human sciences, it seems implausible in the natural sciences. But what I wish to show is that what the dialogical model assumes in terms of the hermeneutical situation applies in terms of the relation between human beings and nature as much as between human beings. Gadamer attempts in *Truth and Method* to show the situation and experience which underlies the human sciences. I wish here to attempt to draw out a similar approach to the natural sciences which sees the natural sciences as Gadamer sees the human sciences, as expressions of an underlying structure of relation. I will do this by radicalizing two basic hermeneutical concepts, the hermeneutical circle and the double hermeneutic.

In the background of this radicalisation is Heidegger's attempt to think human history back into nature, or rather *phusis*. What Heidegger wishes to say with his 'retrieval' of the 'original' Greek meaning of *phusis* is that before history and nature, there is for the Greeks the coming into the open of what is. Indeed, Heidegger concludes his essay on the essence of the concept of *phusis* by characterising *phusis* as the concealing disclosure (*verbergene Entbergen*) of being and hence as truth (*aletheia*).[9] If one reads the title of Gadamer's main work (*Truth and Method*) in the light of this move in Heidegger's ontology, then one has a guiding light of how to think further with and against Gadamer on the truth of nature as well as of humankind.

I wish here to take two tacks in my analysis. First, I will explore the extent to which the ethical element can be shown to be fundamental to the hermeneutical situation, and second I shall argue, by recourse to a Heideggerian and pragmatic concept of truth, that we can no longer divorce truth from goodness. First then to the ethical element of hermeneutics.

Hermeneutics, language, otherness

The hermeneutical circle and the double hermeneutic are in the work of Gadamer and Ricoeur, inseparable. One defines the other. More precisely the hermeneutical circle is based on and assumes the double hermeneutic. The double hermeneutic is heavily influenced by the German idealist concept of self-consciousness. What characterises the double hermeneutic is that the object of study is, in Taylor's term, a self-interpreting animal. This becomes significant in two ways: first the interpreter must take into account the self-interpretations of the objects (persons) of interpretation and second the interpreter's interpretation of these persons, if it enters the public domain, can influence the self-interpretation of these persons. It is not difficult to see how this factor contributes to the historicization of the object of research. This same process does not occur in the case of the natural sciences, or so the argument runs.

This is crucial for the hermeneutical circle, as it is generally conceived, because what characterises it in the sense that Gadamer uses it is that there is a two-way movement of interpretation, from questioner to questioned, where these designations are displaced from one to the other such that there is no dominant voice (interrogator), but rather a play in which "the players are not the subjects of play; instead play merely reaches presentation (*Darstellung*) through the players."[10] For this to be realised, though, the object of interpretation must be a player not an object, must be a 'person' due respect, who speaks with his own voice.

In this sense the hermeneutical circle appears to apply to the sphere of the human sciences exclusively. But the objects of the natural sciences too are in a dynamic circular relation with the investigator, as Kuhn for example has pointed out.[11] But more radically than that, I want to argue that the objects of the natural sciences are involved in an ethical relation. This claim stems from how we conceive of the hermeneutical circle.

In the light of how Gadamer describes the hermeneutical experience, the hermeneutical circle can be said to have an ethical significance. This ethical significance of the hermeneutical circle is two-fold. On the one hand the interpreter is not only a participant in the act of interpretation, but, is in part, constituted by the act of interpretation. In this way the life and ethical practice of the interpreter cannot be divorced from the act of interpretation. This is what is termed the double hermeneutic. But it is ethical in a second way also. The interpreter is called into question in the hermeneutical circle. For this it is necessary that the interpreter allow his deepest prejudices to be exposed by the other. This demands an openness to the other. This is primarily what makes the hermeneutical circle ethical: it is rooted in openness to the other, in the possibility of welcoming the other as the one who will call the assumptions of the questioner into question. For the hermeneutical circle undercuts the claims of the uninvolved observer. The questioner is a participant in the circle, is himself called into question by the circular movement of understanding. A prior condition for this is that the interpreter is willing to listen to the other. In the circle the questioner becomes the questioned, the questioned the questioner. This runs in the face of what Kuhn calls normal science, in which scientists "attempt to force nature into a preformed and relatively inflexible box that the paradigm supplies."[12]

How, however, can this calling into question occur in nature? That another person or the product of a human hand can call us into question is at least plausible. It remains unclear how this is possible with regard to nature. But do we not dwell in response to nature, building our cities on waterways or on mountain tops, making our ships differently in response to rivers or oceans, growing our crops and grazing our animals in response to the soil? In these responses there is a knowldege, a know how, which imposes in the act of responding. But in that response there is also a possible hearing of the question itself: who are we who dwell in cities, who sail ships, who farm? In the response manifest in these activities there is an implicit understanding of nature. This moment of questioning and being called into question lies in every response to nature – not excluding that of science.

The concept of otherness I am employing in regard to the hermeneutical circle is characterised fundamentally, not by the fact that one values the other, but rather that one is not it, one cannot have its horizon. The other has an exteriority in the sense that it enters into relation with me as that which I am not. Its exteriority as such has nothing to do with the will and is always relational. Its otherness resides in its facticity, in its ultimate resistance to incorporation. It is the otherness of the other, which is not breachable. This otherness is not confined to the realm of the human, because otherness, as so defined, can be applied to any realm. It is as such that the 'object' of interpretation enters into the hermeneutical circle, not fundamentally as a being to be valued, but rather as that which cannot be exhausted by the act of interpretation. But this is not the other of the double hermeneutic. I can not enter into the horizon of a rock, but that certainly does not entail that the rock is self-interpreting. What needs to be shown is that the hermeneutical circle is in fact more fundamental than the double hermeneutic and that hence the hermeneutical other is the other which enters into the circle as the inexhaustible object of interpretation.[13]

More fundamental than questions of value are the questions of language and historicity. That it is her place at the intersection of language and historicity that defines the human being, as that being who can constitute herself and constitute her own world, is in effect the claim made by Gadamer and most explicitly found in Taylor. The point of the double hermeneutic is that human beings are uniquely self-constituting, and further that social reality differs from natural reality in the sense that in social reality "language is constitutive of the reality, is essential to it being the kind of reality it is."[14] While this may be so, it does not justify the distinction between the *Natur-* and the *Geisteswissenschaften*.

I do not want to deny that the strength of Taylor's argument concerning human beings as self-interpreting animals, or the centrality of language to this process. I think that his argument is strong and persuasive. However, this difference is not fundamental enough to justify the distinction he wishes to draw between the natural and the human sciences.[15] The central issue here is the nature of language in which self-interpretation and self-constitution occurs.

Levinas argues for the ethical nature of language, but he confines this to discourse, to the human saying with its movement of saying and silence.[16] I wish to extend the ethical nature of language further by calling into question the fundamentality of the constituting factor of language and allowing for the possibility that language is fundamentally a locus of openness and creativity, thus undercutting the claims of the distinctiveness of human beings in terms of creativity (self-constitution) as the 'objects' which must be respected as subjects. The fundamentality of this latter factor can be called into question by recourse to Gadamer himself in his concept of play. What is significant about Gadamer's phenomenology of play is not so much that it transcends the subjectivity of the players, but that it gives precedence, or primordiality, to the *process* of play. As Gadamer states:

> It is the game that is played – it is irrelevant whether or not there is a subject who plays it. The play is the occurrence of the movement as such. Thus we speak of the play of colors and do not mean only that one color plays against another, but that there is one *process* or sight displaying a changing variety of colors.[17]

Gadamer goes on to say that the primordial sense of play is the medial one. This concept of play can fruitfully be conceived in terms of control or rather the lack thereof. The play is most genuine when it is not controlled by the participants. To the extent to which the participants give themselves over to the back and forth of play they loose control. If language is viewed in this way then a middle ground can be found between the objectivist view of language (structuralism) and the view of language as the production of transcendental subjective intentionality (phenomenology). The play of language neither controls the subject nor is controlled by him, rather the process of language happens in the speaking subjects for whom language is larger than themselves, and but for whom language would not come into play.

If this analysis is accepted then the place of the self-constituting subject at the intersection of language and history (which itself brings the subject 'into play') becomes suspect. The human being is that being who is in dialogue (the dialogue which we are). But dialogue itself is a process of play. Unless one subscribes to a private language view then language is only realised in dialogue. But language is essential to self-interpretation and self-constitution. As the play of dialogue supports, rather than being supported by, the self then the process of self-constitution would seem to be displaced the consciousness to the play of language manifest in the hermeneutical circle. The play of dialogue is the phenomenological basis of the hermeneutical circle in the sense that it is only in dialogue that the circular structure of question and answer, prejudice and understanding, is manifest. But if the double hermeneutic, as the characterisation of the role of self-interpretation in the hermeneutical situation, is only possible through language in dialogue, then it becomes a province of the hermeneutical circle rather than the reverse. The hermeneutical circle does not depend on the double hermeneutic, but rather that the double hermeneutic is a product of the hermeneutical circle. If this is so then human self-constitution is a secondary phenomenon in terms of the act of interpretation. It also follows that the characteristics of language and the hermeneutical circle are not grounded in the nature of human beings, but rather that the ethical relation is what makes human self-constitution possible and lies at the basis of the encounter with that which 'calls for understanding,' with all the connotations of this phrase. Hence, the appeal to self-constitution and the double hermeneutic does not mark a fundamental distinction, often characterised as that between *Natur* and *Geist*, which would deny hermeneutical insights legitimacy in the natural sciences.

Pragmatic truth and the good

I want here to repeat the above argument on another level. My concern has been up to this point with the situation and experience of understanding. What has been implied is a re-evaluation of ethics in relation to science. What is true is so only through a linguistic relation, which cannot be abstracted from the ethical.[18] That assumes that one can overcome the division between truth and goodness. If we take the basic tenet of pragmatism seriously, that that which is true is what works, then, ironically, we are on the way to breaking down the distinction between truth and goodness, epistemology and ethics. If what is true is what works, then what is true is what proves true *in practice*. What is true in practice, however, is not what I decide is true, if for no other reason than that my decisions presuppose a space opened up, in which truth is already an issue. This space is the space of practice. My actions are already a response to this situation, which is as such only opened up in the context of my action. The play of concealing and revealing, which Heidegger understands as truth (*aletheia*), does not encounter me as a spectator, but rather in my action as that approach towards what is, which as such becomes a way of truth.[19] But, then, my approach to what is, is not alone a way of truth, but is a way towards the other as that which addresses me, which in a phrase we have encountered before, 'lends itself' towards me, towards my approach. The question of goodness is precisely the question of how we respond to this approach and to this attraction. This same theme comes through in Gadamer's analysis of hermeneutical consciousness. Hermeneutical consciousness is both acted on by, and acts within, history. Interpretation is a matter of practice. The act of interpretation is an act of application. What follows from this is the close link of 'the true' and 'the good.' This is clear in Gadamer but most eloquently expressed by Taylor as follows:

> These sciences [the human sciences] cannot be '*wertfrei*'; they are moral sciences ... their successful prosecution requires a high degree of self-knowledge ... our incapacity to understand is rooted in our self-definitions, hence in what we are.[20]

This linkage of the true and the good is best conceived not in moralistic terms, but as an expression of the fact that there is no real distinction between the knowing and the living self. Knowledge is the actualization of being-in-the-world. What this means is that what we know becomes a part of what we are. The prejudice that this applies only in the case of the human sciences is a deep one and is implicit in the above quote from Taylor. It is a prejudice, however, which, I hope, I have shown to be one which goes against the very logic of hermeneutics itself.

The pragmatist account of the linkage of truth and practice is neatly summed up in Richard Rorty's concept of edification.[21] Rorty uses the term 'edification' to translate Gadamer's use of the term '*Bildung*.'[22] For him edification, not theoretical truth, is the goal of thinking. Edification is achieved through discourse, through the creative redescriptions we make of ourselves. Thus the concept of edification cuts across the fact-value gap. As Rorty puts it, "to use one set of true sentences to describe ourselves is already to choose an attitude toward ourselves." He goes on to criticise the fact-value gap as forcing us "to pretend that we can split ourselves into knowers of true sentences on the one hand and choosers of lives or actions or works of art on the other."[23] The important consequence of Rorty's analysis is that all our discourses, including those of the *Naturwissenschaften*, constitute who we are and are significant for just this reason.

To say then that all our sciences are ethical is to claim that their truths are entwined in their practices, which are parts of discourses which constitute us as we are. In the case of the natural sciences this discourse expresses a certain concept of the relation of man and nature. This concept is underpinned by a goal enunciated by Descartes: "we can use this knowledge [of nature] ... [to] make ourselves lords and masters of nature."[24] This discourse of domination of nature cannot be displaced as somehow unrelated to what it is to be human in this society, nor, if the very language and structure of a discourses relation to what is is ethical, can it escape ethical critique.

198

Conclusion

The main task of a postmodern philosophy of science is to conceive of science in terms of the ethical, ontological and epistemological frameworks of society and to subject it to critique on this basis. I have argued in this paper that one of the ways of doing this is to re-examine the assumptions regarding the relationship between human beings and nature which underlie natural science. The importance of this task resides in the fact that if we as a society are defined by our discourses, then we are defined as much by the discourse of the natural sciences as by those of the human sciences. The natural sciences express our relation to the natural environment. But they are not alone in expressing this relation. Poetry, for example, constitutes another mode of expression of this relation. Poetry, indeed, supplies us with another view from which to encounter science. A postmodern philosophy of science, at least as I conceive it, would be one which would bring into play many different discourses regarding nature, without privileging any one. This in turn would legitimate a hermeneutical approach to nature, i.e. an approach which brings interpretations of nature into play and conceives of the natural sciences as one cluster of interpretations among others. This, I have argued, would involve a conception of nature as other, with which our relation is ethical and not merely epistemological. Such a conception, if thought through in the manner I have suggested, would involve a transformation of the philosophy of science into a discipline concerned with science as a human pursuit directed towards our natural environment.[25]

Notes

1.   The question may be asked, why this is relevant to the philosophy of science itself and not rather to environmental ethics, or some other such field. My justification for conceiving of it as a problem for the philosophy of science is that a postmodern philosophy of science must rest on a postmodern concept of philosophy. What that concept might be I have not the place to explore here, but one essential characteristic would be a suspicion of compartmentalisation. Of course, the very expression 'philosophy of science' indicates a compartmentalisation of sorts, but I would draw the distinction in this way: postmodern philosophy should see the specific in its interrelations, which disrupt and render provisional all divisions. This means that it can start from a particular phänomenon such as science, but then be free to see how it relates to what is around it, to its cultural context.

2. Gadamer: "*Wahrheit in den Geisteswissenschaften,*" *Kleine Schriften*, p. 39. In the Foreword to the second (1990) edition of *Truth and Method*, Weinsheimer, J. and Mrshall, D.G. (trans.), Crossroad, New York (henceforth TM), from the original (1965) *Wahrheit und Methode*, Mohr, Tübingen (henceforth WM), Gadamer, commenting on his phrase 'being that can be understood is language,' states, "it [the phrase in question] does not intend total mastery over being by the one who understands but, on the contrary, that being is not experienced where something can be constructed by us and is to that extent conceived; it is experienced where what is happening can merely be understood." TM pp. xxxv-xxxvi, WM p. xxi.

3. In the foreword to the second edition of TM Gadamer states: "My real concern was and is philosophic: not what we do and what we ought to do, but what happens to us over and above our wanting and doing." TM p. xxviii; WM p. xiv.

4. While implicit in my argument is the assumption that my thesis concerning the hermeneutical approach to the natural sciences is in line with the spirit of the work of Gadamer and Ricoeur, both see the application of hermeneutics only within the field of the human sciences. Both relate the concepts of the hermeneutical circle and the double hermeneutic to the human sciences exclusively. The same holds for the concept of interpretation and understanding in both Gadamer and Ricoeur's work. Gadamer's three stage process of understanding, interpretation and application is tied closely to the human sciences. What ties these three stages together and gives them their unity is the attitude of openness to the meaning of texts or other human artifacts. For Gadamer in TM, while the human sciences are constituted by historically situated interests, and their themes of research are actually constituted by the motivation of the enquiry, the object of the natural sciences can be described idealiter as what would be known in the perfect knowledge of nature. This distinguishes the two sciences in terms of the possibility, if only as a regulative idea, of perfect knowledge (cf. TM p. 285; WM p. 269). In a note added three decades after TM was first published, Gadamer acknowledges that in this passage his division of the human and natural sciences is too undifferentiated. But I think it is reasonable to assume that this retraction is motivated more by methodological considerations than by ontological ones. Even Gadamer's later works, such as the articles brought together in the collection (1981), *Reason in the Age of Science,* MIT Press, Cambridge, still assume the ontological

distinction in much the same terms of moral difference characteristic of TM. The problem Gadamer addresses is the problem of the place of practical reason in a technological age. It is the technologisation of society, not the place of nature, that is his concern. While Ricoeur attempts to reintegrate a concept of method into hermeneutics, he does so explicitly in terms of the human sciences, leaving for the most part unquestioned the *Natur/Geist* distinction. Cf. Ricoeur, "The model of the Text" in Thompson, J.B. (ed.), (1984), *Hermeneutics and the Human Sciences*, Cambridge University Press, Cambridge, pp. 215-221.What is absent from these analyses is the possibility that the objects of the natural sciences may warrant the moral attitude Gadamer, and implicitly Ricoeur, reserves for the objects of the human sciences.

5. This characterisation raises the question as to how 'tradition' is being used in this context.This is a major problem in interpreting Gadamer and I will side step it here. What interests me in the context of this paper is that the relation with tradition is irreducibly dialogical – the 'object' of understanding is a thou – and historical – the 'object' effects us pre-reflectively (*wirkungsgeschichtliches Bewußtsein*). This 'traditional' nature of both the object and the situation of the act of understanding is possibly made clearer by the term Gadamer uses which is translated as 'tradition' *Überlieferung*, which literally means to deliver (*liefern*) over (*über*). For a clear and concise explanation of tradition in (Gadamerian) hermeneutics see Mitscherling, J., (1989), "Philosophical hermeneutics and 'The Tradition'," *Man and World* 22:247-250.

6. TM p. 358; M p. 340, emphasis mine.

7. Ibid p. 361; p. 343.

8. Ibid., p. 362; p. 344.

9. Cf. Heidegger. M. (1978), "*Vom Wesen und Begriff der physis in Aristoteles, Physik B,1*" in *Wegmarken*, Klostermann, Frankfurt a. M., pp. 237-299.

10. TM p. 103; WM p. 98.

11. Cf. Kuhn, T. (1970), *The Structure of Scientific Revolutions*, University of Chicago Press, Chicago.

12. Ibid., p. 24.

13.   The inexhaustibility of the other is crucial because it is precisely this inability to be described *idealiter* which for Gadamer characterises the participant in dialogue.

14.   Taylor, C. (1979), "Interpretation and the Science of Man", in Rabinow, P. and Sullivan, W.M. (eds.), *Interpretative Social Science A Reader*, University of California Press, Berkeley, p. 45.

15.   Cf. ibid., pp. 25-72.

16.   Cf. Levinas, E. (1969), *Totality and Infinity*, Lingis, A. (trans.), Duquesne University Press, Pittsburgh, pp.204-212; for the original French, see (1984) *Totalité et Infini*, Martinus Nijhoff, The Hague, pp.79-87.

17.   TM p. 103; WM p. 99, emphasis mine.

18.   It may here be objected that it is gratuitous to bring truth into this discussion because science (at least natural science) is less concerned with truth than with probability or with what works theoretically and technically. But my concern is to situate science in the broader context of the overall relation to that which science objectivises. Approaching that relation, as I will do in this section, from a Heideggerian view of truth as *aletheia* makes the problem of truth inescapable when examining this relation and, as such, approaches the scientific lack of concern with truth in terms of a *lack* which gives philosophy critical room. For a more detailed discussion of truth in hermeneutics see my (1992) "Truth as a Problem of Hermeneutics: Towards a Hermeneutical Theory of Truth," *Philosophy Today*, pp. 122-130.

19.   Cf. Heidegger on truth: "On the Essence of Truth", Sallis, J. (trans.), in Krell, D. (ed.) (1977), *Martin Heidegger Basic Writings*, Harper and Row, New York, pp. 117-141; for the German original cf. "*Vom Wesen der Wahrheit*," *Wegmarken*, pp. 175-200. This is by no means an uncontested way to interpret Heidegger and I accept that one cannot simply interpret him as placing practice as fundamental to theory, indeed he calls into question this very dichotomy. However, what is relevant here is that truth understood in a Heideggerian manner undercuts any attempt to make truth a purely theoretical matter.

20.   Taylor, p. 71.

21. Rorty, R. (1980), *Philosophy and the Mirror of Nature*, Basil Blackwell, London, pp. 357-389.

22. It is not relevant here to examine the appropriateness of this translation or indeed the value of Rorty's interpretation of Gadamer generally.

23. Rorty, p. 364.

24. Descartes, R., "The Discourse on Method," Stoothoff, R. (trans.), in Cottingham, Stoothoff and Murdoch (eds.) (1985), *The Philosophical Writings of Descartes,* Cambridge University Press, Cambridge, pp.142-3; cf. the French original "Discours de la Methode", *Ouevres de Descartes* vol. 6, Adam, C. and Tannery, P. (eds.), (1973), J. Vrin, Paris, 1973, pp. 61-2.

25. I would like to thank Prof. Gary Madison (McMaster University), Prof. Jeff Mitscherling (University of Guelph) and Prof. Barbara Tuchanska (University of Lódz) for their comments and criticisms of earlier drafts of this article. I also benefited greatly from discussing the content of this article during the conference on Postmodern Philosophy of Science in Dubrovnik.

12 Changing the Subject: A Metaphilosophical Digression

Neil Gascoigne

Strictly speaking, this story should not be written at all. To write it or to tell it is to spoil it. This is because the man who had the strange experience we are going to talk about never mentioned it to anybody, and the fact that he kept his secret and sealed it up completely in his memory is the whole point of the story. Thus we must admit that handicap at the beginning – that it is absurd for us to tell the story, absurd for anybody to listen to it and unthinkable that anybody should believe it.

– Flann O'Brien, *John Duffy's Brother*

I wrote down his statement and immediately after it the sentence. Then I had the man put in chains. It was all very simple. Had I first called the man in and questioned him it would only have led to confusion. He would have lied; when I succeeded in refuting the lies he would have told fresh ones in their stead; and so on.

– Franz Kafka, *In the Penal Colony*

I

In his original introduction to the collection of essays that eponymously honours the 'linguistic turn,' Rorty celebrates its appeal thus:

Linguistic philosophy, over the last thirty years, has succeeded in putting the entire philosophical tradition, from Parmenides through Descartes and Hume to Bradley and Whitehead, on the defensive. It has done so by a careful and thorough scrutiny of the ways in which traditional philosophers have used language in the formulation of their problems. This achievement is sufficient to place this period among the great ages of the history of philosophy. (Rorty, 1967, 33)

Twenty-five years later, he finds himself "startled, embarassed, and amused" to reread the above. "That last sentence" in particular, he writes, now strikes him "as merely the attempt of a thirty-three year old philosopher to convince himself that he had had the luck to born at the right time – to persuade himself that the disciplinary matrix in which he happened to find himself ... was more than just one more philosophical school, one more tempest in an academic teapot." (Rorty, 1990, 1)

One might conclude that the only difference between the *thirty*-something Rorty and the *sixty*-something Rorty is that the latter no longer thinks that a period of greatness in philosophy is any great shakes at all; that a sectarian Hegelianism has merely given way to a secular Hegelianism. What is certainly the case is that Rorty's three introductions to this collection – a second was written in 1975 – chart a precipitous decline in the faith that linguistic philosophy could remain aloof in its metaphilosophical deliberations on the status or reality of this or that philosophical problem.

This is, I think, a great shame; not because there are grounds for renewed faith in the project of linguistic philosophy, but because there is still room for metaphilosophical deliberation on the status of philosophical problems as part of a naturalistic, anti-metaphysical strategy. It is therefore unfortunate that the fortunes of the later have been overidentified with the those of the former. Unfortunate, but not perhaps, with the benefit of hindsight too surprising when one considers that Carnap, for example, held that he could dispose of, say, the problem of scepticism by making the stipulation that it was, like all philosophical problems, an external one, not an internal one; a problem that, as he writes in 'Empiricism, Semantics, and Ontology,' poses a question about the very framework of our beliefs, and does not therefore involve belief in the strict sense at all.

It seems astonishing that the tail-end of empiricism should wag the dog of German Romanticism in this way, and Carnap's crude, irreflexive metaphilosophy has been severely criticised, not least by Stroud in his recent attempt to revive interest in the *philosophical* significance of scepticism. The main charge of having dragged metaphilosophy into the gutter of linguistic philosophy must be reserved for a more sophisticated neo-Kantianism than Carnap's, however: that of Peter Strawson and the whole tradition of Wittgensteinian revisionism that has surfaced in Oxford in his wake. This is responsible for having directed metaphilosophy away from the therapeutic, anti-

206

metaphysical promise of a naturalistically construed version, and back towards its Kantian roots. It is thus deeply implicated in the backlash against naturalism that is to be found in the work of Barry Stroud, Thomas Nagel, Stanley Cavell, Mary Midgely and others; philosophers who seek succour in the very intractability of the problems that linguistic philosophy originally sought to dispose of hygeinically.

The background against which this paper is set, then, is the claim that the contemporary, naturalistically-inclined metaphilosopher has a more difficult and yet ultimately more interesting job than his linguistically-fixated forebears when it comes to dealing with the status of philosophical problems. In part this involves accelerating a process that is already apparent: analytic philosophy's increasing self-consciousness with respect to the historical constitution of its own theoretical tools. Is it not astonishing to find John McDowell, for example, remarking in the Preface to his recent *Mind and World* that it be conceived "as a prolegomenon to a reading of the Phenomenology" (McDowell, 1994: ix)?! Certainly more surprising than the discovery of David-Hillel Ruben's comment that although posesed of: "a very great philosophical mind ... McDowell displays an unfortunate philosophical style in these lectures, which is more akin to that of the post-Kantian German philosophers of the 19th century than to that of recent analytic philosophers." (Ruben, 27)

In part, then, something of a return to Hegel; but a return that is leavened with irony and sensitive to the moves that are permissible in the 'disciplinary matrix'. If Davidson is correct when he observes from the deck of his sloop that there has been "a sea-change in contemporary philosophical thought–a change so profound that we may not recognize that it is coming," (Davidson, 1989, 159) then we metaphilosophical naturalists can proudly identify ourselves with the young Rorty's enthusiasm for the significance of the present, whilst embracing the older Rorty's warning that the sloop we are attempting to navigate is tossing around on what was, and will become again, rather stale tea.

II

The problem with which this paper is concerned is the problem of consciousness, and anyone taking more than a passing interest in the 'subject' will agree that in philosophy of mind the debate about consciousness can be seen to constitute something of a philosophical impasse: even for those who strive to remain non-partisan. Consider, for example, a notice on Dennett's version of *The Phenomenology, Consciousness Explained*, in which David Papineau avers that "[i]n the end Dennett's materialism is probably right, since the ghostly alternatives make even less sense." (Papineau, 29)

For at least one philosopher, then, a materialist victory is close at hand; and yet our reviewer sounds a discordant note, for it seems that Dennett "would be a better apostle if he spent more time diagnosing the seductive pull of inner flames and less trying to blind his readers with cognitive science." (Ibid.) What could better illustrate the philosophical tension. Materialism, or what is usually understood as naturalism, is primarily concerned with what *would* count as an explanation, and yet a champion of cognitive science is grudgingly told that whilst he might be *philosophically* correct, it is not a circus parade of the successes of cognitive science that is going to drag the audience of the Cartesian Theatre onto the stage and make them the actors in their own drama. It is not by a display of the empirical evidence for this position that he is held likely to "convert many believers in mental ghosts," (Ibid.) but by diagnosing their *Sickness unto Death*!

But what form of hortation, one wonders, does Papineau envisage? Surely the requested diagnosis would *not* succeed, were it to be cast in cognitive terms, for if the evidence garnered by scientists has singularly failed to convince at the empirical level, there is little reason to believe that a meta-cognitive story – one addressing itself to the somatic aetiology of the *philosophical* sickness – will fare any better. This will certainly be the case from the perspective of someone like Thomas Nagel, tied to the anti-verificationist intuition that scientific evidence "is bound to leave undescribed the irreducibly subjective character of conscious mental processes." (Nagel, 1986, 7) Where Dennett, is wont to bemoan Nagel's preference for "his mystification to the demystifying efforts of others," (Dennett, 1987, 5) Nagel excoriates Dennett for his "Procrustean conception of scientific narrative," (Nagel, 1991) on the one hand, charges of myth-making and obfuscation; on the other, scientism and 'reductive rage.'

In the clear absence of any common explanatory framework we have – between Dennett and Nagel – something of a *meta*philosophical disagreement. It can be characterised in a number of ways, one of which would be to take Richard Rorty's recent formulation that it represents perhaps the final stage of the 'Battle over Intrinsicality', grounded in whether you believe it possible or not "to defend the claim that there are intrinsic, non-relational features of objects", knowledge of which "is not the same as knowledge of how to use the words which one employs to describe those features." (Rorty, 1993, 186)

Although I happily concur with this analysis, I think that the dispute marks its presence in a way which suggests an alternative strategy to what Rorty refers to in the same piece as his own "lofty metaphilosophical rhetoric" (Ibid., 198) and that is what I want to explore in this paper. Let us therefore consider some of the characteristics of the debate: First of all, we have open hostility to the very idea that cognitive science, or any similar form of inquiry, provides an appropriate context for the explanation of the phenomena of consciousness. As Colin McGinn writes, "Consciousness may be one of the subjects that our biology has not equipped us to understand." (McGinn, 19) It is not the scientists themselves, however, who, speaking with the 'vulgar,' are charged with what Mary Midgley calls 'Reductive Megalomania' and so a metaphilosophical account of the disagreement which is sensitive to this accusation will be attracted to the possibility if problematising the way in which the materialist draws upon the 'success' of the sciences in support of his attempt to replace, as Dennett says, "one family of metaphors and images" of the mind "with another." (Dennett, 1991a, 455) I do not mean to suggest by this that *meta*philosophy can retain a position of privilege with respect to scientific inquiry – we cannot, after all, be Arthur Fine[1] about ontology without seeing something a little Stanley Fishy about theory; but it is perfectly acceptable for the metaphilosopher to examine how the significance of this inquiry is appropriated by (other) philosophers and the implications they derive – the new metaphors, if you will – from it.

The uses and abuses of cognitive science is one consideration for the metaphilosopher then – let us call it the external consideration. But one might agree with McDowell when he says that Rorty's debunking of philosophical problems like scepticism and consciousness fails because such an "external approach can easily leave the philosophical questions looking as if they *ought* to be good ones" thus resulting in "continuing philosophical discomfort, not an exorcism of philosophy." (McDowell, 142, fn. 17) It therefore seems entirely appropriate to attempt to 'do justice to' such 'intuitions' by offering a *normative* supplement when confronted with the hostility of the exchanges between those for and against naturalistics account of consciousness.

In addition to the 'externalist' approach, then, there might be some mileage in Dennett's own Rylean[2] submission that the mystification of consciousness embodies a fear that somehow, without it, moral agency would collapse. Viewed *in this light*, Papineau's moderately reproachful attitude to materialism's *philosophical* success registers a mere ripple of disquiet when compared to the lashing waves generated by the existential sensibility of a Nagel or a Midgely. But its very measuredness is instructive; for what manner of narrative could conceivably satisfy the moral concerns which underlie this 'seductive pull' of inner flames? The question suggests an 'internal' strategy in which the metaphilosopher attempts to do justice to the view that Nagel's 'irreducibly subjective character' implies the existence of a subject whose moral status becomes dubitable when wrested from a particular context – one in which it has a 'point of view,' for example – and which, as a result, generates moral doubts about the whole physicalist enterprise.

In the next section I shall therefore sketch a philosophical framework within which a moral expropriation of this subject might be perceived to be occurring when the intrinsic mysteriousness of consciousness is challenged. Before abandoning our present lyrical and lofty perspective, however, one might think of this existential fear as occupying a distant node on the web of the 'great world-spider' of belief and desire; one which *appears* to the antagonists to be isolated from the conceptual epicentre of the debate, but which is nevertheless 'excited,' let us say, by the causal power of certain metaphorical[3] redescriptions of the mind and gives rise to the ungovernable 'intuition' that third-person descriptions not so much *leave* something out, as *stamp* it out.

III

Consider, then, the kind of philosophical account that once held philosophers captive: one in which Kant is seen to contrive a delicate balance between, on the one hand, the demands of a buoyant empirical science and, on the other, those of securring moral agency. In handing over the empirical domain to the former, the phenomenal 'I' is propelled into *a priori* experience in Space and Time by a noumenal subject that remains transcendentally aloof. The 'me,' the empirical subject, that determination of the forces of nature, is doomed to represent to itself, temporally, in thought, an 'I' that is forever unknowable, and yet acknowledged to be the transcendental condition of my 'being' me. The purpose of this unknowable self is to universalise the subject *qua* undetermined agent. It makes possible a universal ethics in which what is known of Reason is an unconditioned categorical imperative: if your will be imposed on the world in which *you* exist as an empirical being then your will be done.[4] Assert, then, your*self* in the knowledge that you are fulfilling the will of the universal Rational subject to which other empirical selves will respond. Thus did Kant find "it necessary to deny *knowledge*, in order to make room for *faith.*" (Kant, Bxxx)

The problem arises when, with the neo-Kantian interpretation of the first *Critique*, the empirical subject as a constituted object of experience is destined to become an object of theoretical knowledge, the understanding of which lies within the limits of actual or possible, that is to say, theoretically verifiable experience. In this quasi-positivistic reading, the transcendental subject disappears, and with it the notion of a humanistically centred moral agent. So perhaps this gives us a possible approach to our disagreement over the status of consciousness; perhaps it is the case that when Dennett and others attack the notion of the subject being self-transparent in some way – not being aware of the limited acuity of, say, peripheral vision or of pain behaviour – Nagel *et al* see this *not* as a limitation of the *phenomenal subject*, but as an *assault* on the moral authority of the *noumenal subject*. In other words, as an attempt to illegitimately extend the domain of knowledge to that, the existence of which, requires faith, and which for Nagel is the unknowable condition of the very possibility of objectivite knowledge.

Now clearly the story isn't going to be quite as simple as that. After all, no philosopher gets terribly upset by being told that subjects cannot self-diagnose viral infections. Moreover, under this interpretation we appear to be favouring the reductive approach of something akin to Dennett's Rylean view too strongly. We must therefore add a further dimension to our inquiry, and to do that we need to return with a little ironic distance to the fate of the 'transcendental subject'. Now, in addition to its moral function, this unknowable subject serves another purpose, which suggests that it didn't disappear with positivism, but is implicit in any version of transcendental philosophy, of which positivism, as expressed through the legislative function of the verification principle, was one.[5] Consider, then, the *author* of the text. In the case of Kant, he stands in a relation to the *Critique* analogous to that of the noumenal subject to the phenomenal subject: as the very condition of its possibility. He is the one who can deduce the ineluctable categories and identify the Moral Law; the one, as it were, to whom Reason whispers when he asks the immortal *quaestio juris*. It is Kant, the author, who stands at the very limit of the analysis and determines what the subject is: what balance is struck between determination and purposiveness.

What is notable about this 'author function', I suggest, is that it is both *apodictic* and *aporetic*. To develop the point consider the case of a rather näive and self-conscious attempt to obviate its reflexive implications – implications that were implicit in my cursory dismissal of Carnap's metaphilosophical distinction between 'internal' questions and 'external' questions. In the *Tractatus*, Wittgenstein attempts to solve the riddle that while propositions correspond to facts in a relation of 'picturing,' the propositions of the *Tractatus* itself are literally nonsensical, since they imply a perspective, outside language and the world, that would be necessary in order for one to *say* anything about that relationship. As Bernard Williams notes in his influential piece on Wittgenstein's idealism, the Tractarian subject is the "metaphysical or philosophical self," (Williams, 145) which lies at "the limits of my world." (Wittgenstein, 1922, 5.62) The presence of this can be seen to constitute Wittgenstein's attempt to come to terms with the self-referential instability of the Kantian categories, and the recognition that, as David Pears notes, they "cannot be expressed in sentences because they are the pre-conditions of saying anything." (Pears, I: 147) All one can do, according to this formulation, is *show* them: point towards the limits of language. And what lies there, at the limit, is that philosophical self – the 'my' – which is not a part of the world and yet which, like Nagel's mystical finitude, is its grounding.

If we extend the idea of the 'author-function' I related to Kant's metacritical 'I' to Wittgenstein's self-diagnosis, we can perhaps equate it with this *Tractarian* 'philosophical self,' for the limit that cannot be known is still the limit from which the author functions. We are apparently being *shown* that limit from *within* the domain of the empirical, but it is in fact the dais from which he has cast his ladder, and against which ours leans as we climb, for ultimately we will come to recognise the truth about the world that the *Tractatus performs*, and to which Wittgenstein, in his authorial guise, is always-already privy. The situation is thus akin to other forms of structuralism, where the dissolved 'knowing,' or 'meaning-giving,' or 'morally autonomous' subject is re-posed at the limits of the structure in the form of the structuralist/author who is not located within it, and whose knowledge, or perhaps presence, thereby *pervades* it.

Since this singular epistemological perspective circumscribes reality, we are led to the first of three points, the metaphilosophical significance of which will become clearer as we proceed through subsequent sections, but which for the time-being have rather gnomic formulations: (1) that "solipsism, when its implications are followed out strictly, coincides with pure realism." (Wittgenstein, 1922, 5.64) Combined with the suggestion from the previous section regarding the need for a consideration of a normative dimension if we are to diagnose our disagreement, this leads to the other two points: (2) that the 'philosophical self' can be identified with the Kantian moral subject: "the human soul with which psychology deals" (Wittgenstein, 1922, 5.641) "do[es] not relate to an 'I' *in* the world, and hence we cannot conceive of it as a matter of empirical investigation" (Williams, 146); and (3) that the conditions of the possibility of knowledge cannot be expressed, but lie at the limits constitutive of the *author's* philosophical self. With these in mind, let us consider a passage from Consciousness Explained[6] in which Dennett introduces the idea of heterophenomenology, a key term in his account of the 'subject.'

IV

If one doesn't share the Nagelian intuition regarding consciousness then one might be led to *question* what status to attribute to the term. As a good analytic philosopher one is going to have to either acknowledge that it is in fact referring, or cohering, in some sense, or regard it as hopelessly vague and empirically abstruse, best shunted off to a home for the semantically redundant.[7] Since we *tend* to think it is meaningful because we *tell* one another we are conscious, an alternative is to suggest that one might learn something about its nature by listening to others *speak*. But, cries Dennett, isn't this exactly what psychologists do? Better still, unlike those dreadfully ambiguous interactions which take place in the *Lebenswelt*, psychologists construct objectively describable and delimited contexts in which a resolutely third-person perspective can be maintained, and actors happily answer questionnaires, press buttons and gaze at other 'actors' pretending to be tortured – all at the behest of the wilfully dour and impartial observer-scientist.

This suggests to Dennett a way in which one *might* make good empirical use of the concept of human consciousness. Why not transcribe the agent's utterances. Certainly, there will be some problems – slurs, stutters and spoonerisms to be weeded out – but since the text consists of speech acts and is, as he says, "a product of a process that has an *intentional interpretation*," (Dennett, 1991a, 161; 76) there should be little problem correcting these. After all, if someone says 'from light to reft' it is obvious that they meant to say ... well, as Dennett says, it's obvious. What remains, after this process, is consequently described as an "error-corrected ... purified" (Ibid., 161; 760) textual rendering of the subject's current beliefs and opinions which can subsequently be submitted to that branch of literary criticism known as hermeneutics.

When we read a novel we flesh out the details of a cosy, armchair counterfactual, passing well beyond the *actual* contents of the text. We *know* that along with Sherlock Holmes there are no jet-planes and philosophers of a linguistic *bent* – no let's say *turn*. I recollect no mention of them, but we know that syphilis and piano tuners can be read between the lines of those smoggy London streets. Furthermore, says Dennett, rather more *sans* Quine than one would expect, we can know the *story* of, say, Madame Bovary, without reading the book – indeed Dennett claims to be able to write a passing term paper on the subject on the basis of having seen the BBC T.V. adaptation. The 'key assumption' here is, as he tells us, that "we can describe *what* is represent*ed* in *Madame Bovary* independently of *how* the represent*ing* is accomplished." (Ibid., 165; 80) If, in this *right* – sorry *light!* – we view the actor as author, a whole world of objective facts lying behind her conscious intentions can be added to those that are explicitly stated in her text; and it is this world that is said to surround the *real* subject. The 'I' is the – possibly imaginary – locus of this objective outsider's view of the subject's consciousness, termed by Dennett the heterophenomenology of the subject in contrast with the subject's own autophenomenological self-conception. Denying himself a terminal exclamation mark – though with the consolation of some spirited capitalisation – Dennett goes so far as to suggest that this "may even be a case of the Humanities lending a Helping Hand to Science"! (Ibid., 164)

In response to Dennett's authorial de-contextualisation of the subject which is masked, somewhat, by this folksy talk of the 'empirical study of human consciousness,' we might want to consider a gently deconstructive line, which might go something like this: "yes, but Professor Dennett, you talk about these 'facts' of the subject's possible world which lie behind and beyond their 'text,' but these are just a result of that worldly and promiscuous interplay of texts. You may feel, as a self-annointed 'metaphysical minimalist,' that this speech, *wrighted* as text, must admit of an objective interpretation,[8] but the only context which is going to be meaningful, once one acknowledges this semantic leakage, is the Davidsonian 'given' of the *One World View* that bounds – and binds! – *all* texts and ensures that the majority of *all* our beliefs are true.[9] But this does not therein deliver unto a *philosophically* sanctioned cognitive science a position of epistemological privilege in respect of them. You have played with the fire of hermeneutics and should accept that you now find yourself in a leaky intertextual world of slippery significations and ecstatic differences."

Since Dennett has displayed at least a passing interest in literary theory, however, it seems only appropriate that we should develop this approach in a spirit conducive to our more general metaphilosophical diagnosis. Let me recount, then, a story *within* this story.

V

Rummaging through a collection of books recently, I discovered what appeared to be a *journal*. The cover, and many of the (numbered) pages, were missing but despite the absence of an identifying surname it was clearly the work of an author called Édouard who was quite probably French. I read it with interest and soon deduced that he was, at the time, working on a novel, with which I was not familiar, provisionally titled *The Counterfeiters*. In the course of my perusal I managed to work out that Édouard had a nephew, George, suspected by his mother of stealing 100 Francs – internal evidence that it was, indeed, a translation from the French. She had informed Édouard of her suspicion and furthermore, he tells us, asked him to talk to George about the affair. Édouard subsequently confides his discovery that George is also involved in the activities of a gang who are circulating counterfeit currency and, moved by his sister's entreaties, decides to act.

In a later entry he describes his attempt to make evident his knowledge of George's criminal enterprise. During the course of an interview, and feeling inadequate to the task of interrogating George directly, Édouard shows him an extract from his planned novel – the only part we find in his *journal*. It consists of a discussion between two characters, Audibert and Hildebrant, in which the former introduces the subject of the crime of a third, Eudolphe. The exchange, Édouard confides to his *journal*, is more or less a transcription of the conversation which had taken place with his sister: "It very rarely happens that I make direct use of what occurs to me in real life," he wrote, "but for once I was able to take advantage of this affair of George's; it was as though my book had been waiting for it, it came so pat; I hardly had to alter one or two details." (317) Édouard confesses to the reader that in the light of subsequent events his idea

214

now seemed absurd, but adds that *in his novel* "it is precisely by a similar reading that I thought of giving the youngest of my heroes a warning. I wanted to know what George's reaction would be; I hoped it might instruct me ... and even as to the value of what I had written."

In reading the extract, then, George is meant to understand that the fictional conversation is about him*self* and Édouard has hopes, through that recognition, of empirically confirming the psychological hypothesis that will play a role in rendering credible an important section of his book. The narrator of the extract from Édouard's novel, *The Counterfeiters*, informs us that Audibert is recounting the story of Eudolphe "without naming him and altering the circumstances so that Hildebrant should not recognize him." (318) Édouard, falling prey to his novelistic – the *writer's* – need to control his characters, is using a *text* to communicate what he cannot control in a face-to-face confrontation with another. So how does the seduction work? In showing George the extract, Édouard is offering him the opportunity to become his ideal reader, one beyond the ken of literary critics and authors but not, I fear – from what we have seen – Dennett. If he is successful, both the (performative) artifice *of*, and the truth *in*, the text would be realised. George is being invited to play the rôle of both Hildebrant *and* Eudolphe – both in the third person. By gaining access to the narrator's level of description – Édouard's by implication, whose silent presence pervades the context in which the reading is taking place and which determines its meaning – George is made present in *The Counterfeiters* by identification with Hildebrant, *at the expense* of his absence as Eudolphe, his true cipher. He is being offered the possibility of aspiring to Édouard's magisterial perspective, but only if he surrenders to his claim to be the supreme author of the situation, thus rendering George a character in George's own life, as well as in Édouard's fiction: that is to say at the expense of power for him*self*, his *own* authorship.

The attraction is to be immortalised by allowing one's destiny – planned out in advance, determined by a higher authority, and therefore *free of responsibility* – to be idealised. That is to say, *to surrender choice* in the interest of idealisation of form by another. Just as literary characters, those protean incomplete creatures, are created when one *structures* – let's say *contextualises* for consistency's sake – one's sudden insights, intuitions, imaginative leaps, Édouard is attempting to make George – the George he knows descriptively through the appropriation he has made *of his history* in the detail of his *journal* – *subject* to that very process. Édouard can authenticate his own authorial voice – exercise the Will to Truth seen as the ideal of Art – by imposing it on George. How can George resist?

Now the parallel with Dennett will not, I hope, seem too far-fetched. The great problem people present for the experimenter, Dennett tells us, is that they often tell others what they think they want to hear. To combat this Dennett suggests that "we do what we can ... to put them in a situation where given the *desires* we have inculcated in them, they will have no better option than to try to say what they in fact *believe*." (Dennett, 1991a, 162; 77) If this behaviouristic preparation of the subject strikes you as troubling, we should remember that for Dennett "people are often just wrong about what they are doing and why they are doing it." We have, he adds, citing Gunderson with pleasure, "*underprivileged access* to our own mentation."(Ibid., 173; 94)[10] Human subjects are notorious dissemblers and incorrigible beasts, who would rather fictionalise their lives than admit they cannot present an internally coherent account of their beliefs to the third-(non)-person observer who has gone to such trouble to prepare them. We, the unwitting authors of our own texts, are denied knowledge of the real beliefs which give them status as truths about us.

Fortunately what *we* are denied, the impartial experimenter can access. He can present the idealisation of our own mentation. The heterophenomenological view *is* the objective outsider's view of that subject's consciousness. It is of course open to the scrutiny of the author-subject but, as we have seen, she generally lacks the resources to "permit a direct, factual, unmetaphorical recording of the events [s]he wishes to recount." Being locked into a self-deceptive fiction, these may have to be "drastically reinterpreted ... if necessary over the author's anguished protests ... to reveal a true tale, about real people and real events ... in such a way that its terms can then be seen to refer – in genuine nonfictional reference." [168; 84] The heterophenomenologist appropriates the subject as object, subverting the *knowing* subject and legitimating her narrative only insofar as its articulations are in accordance with the experimenter's idealised account.

VI

Now, I was moderately satisfied with my interpretation of Édouard's ploy but, rather like the 'reader' in Calvino's *If On a Winter's Night a Traveller*, was intent upon discovering who this novelist Édouard *really* was and what, *in fact*, happened to George. The extracts I had in my possession just didn't cohere sufficiently to satisfy my modernist tastes, and yet I was a detective without a lead. Then serendipity intervened. One day, walking through a market – not alas the Alcano at Toledo,[11] but a more homely affair in Cambridge – I happened upon a book by André Gide entitled *Les Faux-Monnayeurs*. Significantly, the only text he consented to call a *novel*, it recounts events in the lives of several characters in Paris some time during the first decade of the twentieth century. The main character is Édouard, a novelist whose latest work-in-progress is called *The Counterfeiters*, and whose presence in the narrative is supplemented by several extracts from his *journal*. In this he discusses his relationships with the various other *characters* of *Les Faux-Monnayeurs*, one of whom, his nephew George, is suspected of, among other things, the crime which gives *both* 'novels' their theme and title.

So it was Gide, then, absent from the *journal*, who was the condition of possibility of the text I had read. I later discovered that in Gide's own *journal* the following event is recalled in July 1914:

> I stupidly left at Cuverville, in the excitement of leaving, the little notebook just like this one, only four days old, but in which I had written last night, or this very morning, some rather sombre reflections about K. Will the fate we associate with fiction arrange it so that he reads them? I am almost inclined to wish so, if only this would lead to some protest, some salutary reaction on his part. (Gide, 1967, 209)

This expanded context suggests to the reader that Édouard, as a third-person presence in the main narrative of *Les Faux-Monnayeurs*, stands in the same relation to Gide as Audibert to Édouard in *The Counterfeiters*. In an entry for November 1924, Gide writes that "[i]t is certain that if *I*, the novelist, have in me the character of the novelist Édouard, I must have also the novel he is writing." (Ibid.: 383) Moving from the inner narrative, which we compared to Dennett's, to the outer, we now see that Gide is representing the act of representation in his novel, himself present in the role of Édouard, the author of *The Counterfeiters*, despite his absence as the author of *Les Faux-Monnayeurs*. Gide's structuring of his own experience, itself represented in *his journal*, is mirrored in the action of Édouard; and Gide's own seduction of the reader is reflected, made manifest, and therefore undermined by the assault Édouard makes on George's autonomy by in turn presenting him with the text of *his* novel.[12]

Early in *Consciousness Explained* Dennett writes that

> just about every author who has written about consciousness has made what we might call the *first-person-plural presumption*: Whatever mysteries consciousness may hold, *we* (you, gently reader, and I) may speak comfortably together about our mutual acquaintances, the things we both find in our streams of consciousness. And with a few obstreperous exceptions, readers have always gone along with the conspiracy." (Dennett, 1991a, 67)

One thinks here, then, of those who, like Searle, enjoin us "always [to] insist on the first person point of view" (Searle, 451) and attempt to gather us to their standard with 'thought-experiments' involving dupes in rooms, ignorantly shuffling bits of paper around or analysing the colour-experiences of others. In each case being shut-off from the very world we – you, gentle reader, and I – occupy when we are epistemologically off-duty. These, then, are the targets of Dennett's project, what he calls "intuition pumps" (Dennett, 1984, 17) that are tied to the views of the mental he is seeking to displace with his new metaphors of the mind like heterophenomenology. But we must also consider the examples given by third-personists. *Their* characters, orchestrated like puppets, go around acquiring beliefs without apparent justification, and we, the readers, are invited, at one level, to share the author's God's-Eye view.[13] But the author never appears in the text. Like the Tractarian 'I' it lies at its limits, the unknowable condition of possibility. And it is this authorial, solipsistic self that, I suggest, 'coincides with pure realism'. In other words, being absent from the empirical world of the text then, as Wittgenstein writes, it "shrinks to a point without extension and there remains the reality co-ordinated with it." (Wittgenstein, 1922, 5.64) So the world of the text becomes a representation of the 'Real World' and it is to this authority the author appeals when, in our names, he reveals that his characters are in error. I say 'in our names' because the 'author-function' invites the complicity of the reader *qua* subject. Indeed, it idealises the subject insofar as it mediates between the reader and that *aporetic* grounding to which it alone has access. And the 'seductive pull' of that idealisation only takes place if the reader legitimates the author's privileged perspective.

Kant, as we have seen, idealised a *moral* subject, one to which neither he nor we could lay claim *qua* empirical subjects. Only if we acknowledged Kant's authorship – his unique access to a Universal Reason did our moral status become legitimate. Thus the tyranny of the noumenal. Édouard wanted to deny George agency in order to idealise an *aesthetic* subject. One in which the form of the content of George's story – *his* life – was perfected by the author. And yet this formal constraint did not apply to the author himself. He legislated in the name of art – Thus the tyranny of the aesthetic. Finally, with Dennett, we have the idealisation of the *cognitive* subject. His story, as we have seen, is one of confabulating 'authors' and inscrutable psychologists. Like Kant with the *Critique of Pure Reason*, he does not appear in the role of one of those empirical characters to whom, as an author, he is appealing. Privy to the counsel of cognitive reason, in possession of an interpretative framework under*written* by the authority of an objective scientific method, he remains aloof. At the limit of the text. Read *in this way*, he is not a lamb-like, absent, "irremediably narrow-minded and unhistorical philosopher," (Dennett, 1982a, 349) casting a passive and delighted eye over the work of cognitive scientists. As the irreflexive author, his self 'coincides with pure realism': embodying the grounding narrative for the psychologists *manqué*, each of whom becomes a textual manifestation of Dennett's own transcendental perspective. So just as the events in *The*

Counterfeiters are not only based on those 'real' events which take place in *Les Faux-Monnayeurs* but, seeking to idealise, aim to *direct* them, Dennett is attempting to write the lives of his 'authors' and seduce us, his readers, into a simultaneous identification with both the knowing author and the finite, dissembling 'authors.' Thus the tyranny of the empirical.

VII

I began by responding to Papineau's request for a diagnosis of why 'people' were not going to give up 'mental ghosts' by being presented with reams of evidence deriving from cognitive science. The reading of Gide was offered not just as a demonstration of how texts can be re-contextualized to suit the purposes of the author. It was intended to show how a re-description of Dennett's central heterophenomenological metaphor might do justice to the moral-existential disquiet I insisted upon identifying in the responses of Nagel and Papineau. In drawing Kant and Wittgenstein into the story I wanted to make the point that Dennett and Nagel could in fact be read as sharing a common framework – one which rendered their metaphilosophical dispute irresolvable. In line with this, my claim is that there is *no difference* between the thought-experiments of the first-personists and third-personists. In each case one is being directed to identify with both the authorial view – the narrator of the Chinese Room story;[14] Mary the colour scientist's red-blooded teacher on the outside;[15] the heterophenomenologist – *and* the character in the text: the man shuffling bits of paper; Mary herself; the autophenomenologist. According to this, I read 'first-person' as invoking the idealised moral subject we found in Kant: the kind we like to think of ourselves as being when we are not consciously engaged as means to ends; which may well mean just those moments when we are doing philosophy. Similarly, this implies that in holding onto even what Dennett describes as his "mild and intermediate sort of realism," (Dennett, 1991, 29) one can *read* Dennett as simply inverting not Plato, but Kant; or, rather, the transcendental subject Kant bequeathed to one philosophical tradition. Within his half of this undialectic framework, *viewed from the other*, we find that the deployment of these new metaphors for consciousness simply invert the respective roles of the noumenal and phenomenal, the former now displaced from its position of moral authority – carrying the flame of a 'higher truth' about man and in whose name one speaks for all – by the latter. *And yet the Philosophical-authorial voice remains constant.* The point here is that the great polarity of these two 'subjects' – the moral and the empirical – has been translated into a continuum and mapped onto our holistic tapestry of beliefs and desires, so that when Dennett is constructed as the irreflexive author of

Consciousness Explained, his 'new images of the mind' can be seen to emanate waves of existential disquiet by conflating disparate views of the subject and privileging one vocabulary – that of empirical cognitive science – over others. In other words, adopting, now in the name of science, precisely that disembodied and prescriptive authorial perspective which Kant took for himself in the name of a morally authoritative reason.

Have we – I – been fair to Dennett? Or have I indulged in a querulous display of straw-man knocking? I earlier mentioned the view that takes the mystification of consciousness as a reaction to the fear that without it moral agency will collapse. "I am confident," Dennett writes, "that these fears are misguided .. that the widespread acceptance of my vision of consciousness would not have these dire consequences in any case ... And if I had discovered that it would .. I wouldn't have written *this* book." (Dennett, 1991a, 25)

What is striking about this is the extent of the self-identification ('I'; 'me') with the 'objective perspective' and the degree of hubris attached to the moral prognostication. Whatever the intended irony, it is this presumptuous *tone* which irritates, upsets and dismays those *other* first-personists. But how ironic is Dennett's view of heterophenomenology? Is there evidence for seeing Dennett as another more-or-less post-Quinean pragmatist intent on dismantling the Cartesian framework, and whose enthusiasm for science merely causes him, like Quine, to occasionally overstep the mark? Is the intentional stance as he understands it another version of radical interpretation? The simple answer has to be no: a "deviation from normal interpersonal relations is," he informs us, "the price that must be paid for the neutrality a science of consciousness demands." (Dennett, 1991a, 83) His view of consciousness and the role of science is, in fact, highly reminiscent of Freud's. In *Civilisation and Its Discontents*, Freud writes

> Normally, there is nothing of which we are more certain than the feeling of our self, of our own ego. This ego appears to us as something autonomous and unitary, marked off distinctly from everything else. That such an appearance is deceptive, and that on the contrary the ego is continued inwards, without any sharp delimitation, into an unconscious mental entity which we designate the id and for which it serves as a kind of facade-this was a discovery first made by psycho-analysis. (Freud, 12)

It requires no great leap of imagination to see in Dennett's autophenomenology the Freudian ego (self); nor in heterophenomenology the id. Even the suggestion that the two can be brought into harmony with the assistance of the 'objective' cognitive scientist seems curiously reminiscent of the 'deviation from normal relations' which constitutes the therapeutic practise of analysis. There are two other links to be made, however. As is well known,

Freud thought psycho-analysis would eventually link up with neuro-physiology.[16] Dennett holds that heterophenomenology is a "neutral path leading from objective physical science and its insistence on a third-person point of view, to a method of phenomenological description that can (in principle) do justice to the most private and ineffable subjective experiences, while never abandoning the methodological scruples of science." (Dennett, 1991a, 72) Once the heterophenomenological picture has been fleshed out "the way is left open to trade ... heterophenomenological objects ... in for *concreta* if progress in empirical science warrants it." (Ibid.: 95-6) From his position of 'moderate realism,' it will of course be consistent to view consciousness – auto-phenomenology – in terms of a 'limited access' to the heterophenomenological level of 'intentional objects,' ones which are now open to empirical scrutiny. So Dennett must retain a distinction between interpretation and explanation; he must distinguish between the hermeneuticist adopting the 'intentional stance' *qua* radical interpreter and the theorist specifying causes in terms of intentional objects. And for this reason he remains captivated by the inverted Kantianism which morally offends – being so similar to that of – his metaphilosophical antagonists.

I mentioned two links with Freud. The *first* concerned the relationship between 'psychology' and science, and was intended to refer back to the claim that when one attempts to use science to legitimate one's view of the *subject*, one is in effect, from the perspective of the first-personist, changing it: replacing a morally-centred subject with a epistemologically centred one (the noumenal self with the phenomenal self). If one dispenses with the framework shared by Dennett *and* Nagel, one will no longer feel inclined to favour auto-phenomenology over heterophenomenology nor the other way round: for the former reduction can only be effected by a sort of intuitionistic tyranny; and the latter by a reification of the empirical subject in the name of science. If one can resist the inclination towards *either* form of reduction – that is to say, deny the author the wherewithal of what I shall call the 'philosophical stance' – we will have subjects of cognitive science and subjects of 'normal interpersonal relations' and each will, so to speak, potter along with the 'vulgar.'

The *second* relationship refers back to Dennett and objective authorship. The epistemological legitimacy of the whole Freudian programme was based, in part, on Freud's self-analysis. It was this which allowed him to train others to 'know themselves.' *Apodictic* and *aporetic*, it formed the initial act of creation, one which could never be gainsaid, for that would involve an act of resistance on the behalf of the analysand/author-subject, one which, with their complicity, could be 'drastically reinterpreted ... if necessary over the author's anguished protests.' Like Kant's access to Reason, then, Freud had access to a foundational self-analysis; one which placed him outside the context of his own interpretations as one who had harmonised his, let us say, heterophenomenological and autophenomenological worlds. And so of course to Dennett's invocation of the rigorously 'third-person' point of view of 'scientific method'; the one which never locates his confabulations but which allows him to propound *his* 'vision of consciousness' and assure us that if he had discerned any dire threat to our moral personhood he 'wouldn't have written' that particular book. This is just a denial of Dennett's otherwise advocated 'naturalism.' The appeal to scientific method is another example of the 'philosophical stance,' now in its reductive-physicalist guise, whereby one attempt to ground discourse *about* the subject by an appeal to an inapplicable standard. Of course, to misquote Davidson, there is *no such thing* as the subject if it is taken to be what Dennett or Nagel think it. The *real* first-person plural is a fully naturalised and vulgar 'we,' and no moral, aesthetic or empirical subject stands at its limits. Once one appreciates that, the philosophical stance becomes a non-naturalistic view of philosophy which mistakes the creativity of philosopher-authors and a contingent tradition, for what Nagel calls a 'transcendent impulse.'

VIII

So what of George? When we left him, you will remember, he was perusing the part of Édouard's novel, *The Counterfeiters*, in which Hildebrant and Audibert discuss Eudolphe. It ends with the former suggesting to the latter that he write down the conversation they have just had and show it to Eudolphe. After reading this, George – the real subject in the exchange – expresses regret that Édouard had not yet completed the novel and suggests that it would have been interesting to see how Eudolphe had reacted upon reading the notebook! Édouard, dismayed, concludes with the reflection that 'the chapter' would have to be re-written, and that it would have perhaps been better to have spoken with George/Eudolphe. Reality having failed to provide art and author with an idealised subject, however, Édouard loses interest in George. Eudolphe, the true and proper object of his art, will be brought back onto the path of honesty.[17]

References

Carnap, R. (1950), "Empiricism, Semantics, and Ontology," in Rorty (ed.) 1967.
Collingwood, R.G. (1970), *An Autobiography*, Oxford University Press, Oxford.
Davidson, D. (1974), "On the Very Idea of a Conceptual Scheme," in (1984), *Inquiries into Truth and Interpretation*, Clarendon Press, Oxford.
Davidson, D. (1978), "What Metaphors Mean," in (1984), *Inquiries into Truth*

and Interpretation, Clarendon Press, Oxford.

Davidson, D. (1989), "The Myth of the Subjective," in Krausz, M. (ed.), *Relativism: Interpretation and Confrontation*, Notre Dame University Press, Notre Dame.

Dennett, D. (1982), "How To Study Human Consciousness Empirically Or Nothing Comes To Mind," *Synthese* 53.

Dennett, D. (1982a), "Comments on Rorty," *Synthese* 53, pp. 349-56.

Dennett, D. (1984), *Elbow Room: The Varieties of Free Will Worth Wanting*, MIT Press, Cambridge.

Dennett D. (1987) *The Intentional Stance*, MIT Press, Cambridge.

Dennett, D. (1991), "Real Patterns," *The Journal of Philosophy*, vol. 88, No. 1.

Dennett, D. (1991a), *Consciousness Explained*, Allen Lane, London.

Fine, A. (1986), *The Shaky Game: Einstein, Realism and the Quantum Theory*, University of Chicago Press, Chicago.

Fine, A. (1986a), "Unnatural Attitudes: Realist and Instrumentalist Attachments to Science," *Mind* XCV, pp. 149-179.

Förster, E. (1989), "How Are Transcendental Arguments Possible?," in Schaper and Vossenkuhl (eds.), *Reading Kant: New Perspectives on Transcendental Arguments and Critical Philosophy*, Basil Blackwell, Oxford.

Freud, S. (1930), *Civilisation and Its Discontents* (SE 21), Strachey, J. (trans.), W.W.Norton, New York, 1989.

Gide, A. (1966), *The Counterfeiters*, trans. D.Bussy, Harmondsworth: Penguin.

Gide, A. (1967), *Journals* 1889-1949, 'Brien, J.O. (ed., trans. and sel.), Penguin, Harmondsworth.

Jackson, F. (1982), "Epiphenomenal Qualia," *Philosophical Quarterly* 32, pp. 127-36.

Kant, I. (1956), *Critique of Pure Reason*, Smith, N.K. (trans.), Macmillan, London.

McDowell, J. (1994), *Mind and World*, Harvard University Press, Cambridge.

McGinn, C. (1992), "Getting the wiggle into the act," *London Review of Books*, September 10, 18-9.

Midgely, M. (1992), "'Reductive Magalomania," *ms.*

Nagel, T. (1986), *The View from Nowhere*, Oxford University Press, Oxford.

Nagel, T. (1991), "What we have in mind when we say we're thinking," *The Wall Street Journal*, Nov. 7.

Papineau, D. (1992), "Knowing your arms from your elbow," *The Sunday Review*, March 22: 29.

Pears, D. (1988), *The False Prison*, Oxford University Press, Oxford.

Putnam, H. (1981), *Reason, Truth, and History*, Cambridge University Press, Cambridge.

Rorty, R. (1987), "Unfamiliar noises: Hesse and Davidson on metaphor," in *Philosophical Papers* vol. 1, Cambridge: Cambridge University Press 1991.
Rorty, R. (ed.), (1967), *The Linguistic Turn*, University of Chicago Press, London.
Rorty, R. (1990), "Twenty-Five Years After," (*ms.*)
Rorty, R. (1993), "Holism, Intrinsicality, and the Ambition of Transcendence," in Dahlbom, B. (ed.), *Dennett and His Critics*, Blackwell, Oxford.
Ruben, D-H., (1995), "A reconciler of distinction," *Times Higher Education Supplement*, Feb. 3.
Ryle. G. (1949), [1963], *The Concept of Mind*, Penguin, London.
Searle, J. (1980), "Minds, brains and programs," *Behavioral and Brain Sciences*.
Strawson, P.F. (1966), *The Bounds of Sense*, Methuen, London.
Wilkes, K. (1984), "Is Consciousness Important?," *British Journal for the Philosophy of Science* 35, 223-43.
Williams, B. (1974), "Wittgenstein and Idealism," in (1981) *Moral Luck*, Cambridge University Press, Cambridge.
Wittgenstein, L. (1922), *Tractatus Logico-Philosophicus*, Blackwell, Oxford..
Wittgenstein, L. (1953), *Philosophical Investigations*, Anscombe, G.E.M (trans.), Blackwell, Oxford.
Wolff, R.P. (1973), *The Autonomy of Reason*, Harper and Row, New York.
Wolheim, R. (1973), *Freud*, Fontana, London.

Notes

1.   Cf. Fine 1986 and 1986a.

2.   Ryle, Ch. 1; Dennett 1991a, 22-25.

3.   See Davidson 1978; Rorty 1987.

4.   There is of course considerable debate about the identity or otherwise of the 'transcendental self' and the 'moral self.' This is not relevant to my point; the interested reader is referred to Förster, Strawson, and Wolff.

5.   Cf. Collingwood for a similar assertion.

6.   Dennett 1991a. The section with which I am dealing with primarily is derived from Dennett 1982. Page-numbers in the text refer, where possible, to both sources and in historical order.

7.   See Wilkes.

8.   Dennett (1991a) writes that it is "crashingly obvious" that "Holmes was smarter that Watson" and that in such "obviousness lies objectivity."(80) If, by such a claim, he means that the majority of *contemporary* readers would give warrant to such an assertion I agree; but this consensus was not obtained by *imposing* a context of interpretation on the texts: one which the readers in some sense remain outside of. And this is what the parallel with cognitive science would seem to require if one is to arrive at a "*definitive* description."(83) I can see no reason why interpretative practices should not alter in such a way that it becomes fashionable ('crashingly obvious') to read Conan Doyle's texts as, say, ironic reflections on the limits of instrumental rationality.

9.   Cf. Davidson 1974.

10.   The quote from Gunderson is not repeated at Dennett 1991a, but a similar point is made there.

11.   In *Don Quixote*, Chapter IX, Cervantes interrupts a battle between the eponymous hero and a Basque to inform us that at this point the original 'history' upon which he drew ended. He goes on to recount the fortuitous discovery in Toledo of Cide Hamete Benengeli's 'History of Don Quixote de al Mancha' which provides him with the rest of the material for hi(s)tory.

12.   And in which he could be said to be conspiring with Gide in the construction of the 'public' narrative of *Les Faux-Monnayeurs*. Édouard the absent noumenal 'I' to Gide's thing-in-itself creating the empirical space of the novel 'proper.'

13.   Cf. Putnam.

14.   See Searle.

15.   See Jackson.

16.   Cf. Wolheim, Ch. 2.

17.   An early version of this paper was read at the conference on *Postmodern Philosophy of Science* at the InterUniversity Centre in Dubrovnik. My thanks to the following for comments on later drafts: Alison Ainley, Babette E. Babich, Andrew Bowie, Mark Collier, John Forrester, Simon Glynn, Nicholas Jardine, Richard Rorty.

13 Heidegger's Conception of the Technological Imperative: A Critique

Alphonso Lingis

The historical form of the imperative[1]

Martin Heidegger set out to bring to light the history of the specific form of the imperative at work in our theoretical and practical reason.

There is in our history and in the history of our world a certain cast (*Gestell*) which precedes and makes possible both the substantive layout of the field of theoretical perception, that is, of empirical observation, and the predicative structure of our judgments about the phenomenal field. This casting has functioned throughout a certain epoch of history as a destination, an imperative form. All entities we can perceive – and handle and deal with – were from the start cast in a substantive form, of which the modern recasting of the entities given to scientific observation as facts and as objective objects is a derivation. All thought in which we conceive those entities and all the affirmations by which we itemize their properties and relationships are destined to have the forms formal logic has disengaged in order to be cognitive. This imperative casting is the a priori prior to the a prioris of observation and thought.

Rational thought processes are subject to propositional calculus; formal logic disengages its axioms and rules. The axioms of formal logic and the axioms of formalized mathematics, the science of pure forms of quantity, have been shown to be derivable from a set of ultimate axioms, those of mathematical logic. The rules of propositional calculus as well as the rules of mathematical calculus govern the formulation of equivalences, the validity of equations. The syllogisms calculate, for any two or more premises, the equivalent as the conclusion; mathematical equations calculate the equivalent of multiplicities or ratios.

The rules of propositional calculus and the rules of mathematical calculus determine what forms thought must have in order to be valid, what forms multiplicities and sets must have in order to be cognizable. For our empirical science what counts as a report on the universe is cast in mathematical form; the observed is apprehended through measurements. A prior casting of the phenomenal nature as a self-contained system of motion of units of mass related spatiotemporally makes phenomenal patterns lend themselves to being grasped in the forms of calculative thought. There is then a prior casting both of the objects of thought, in calculable forms, and of thought itself, in the rational forms of judgment, that destines the one for the other. The subjection of thought to the universal and the necessary is an aspect of the a priori casting of thought in the form of rational calculus.

The formal mathematical logic of modern times, which exhibits, as deducible from one set of axioms, the system of the rational forms of thought and the system of mathematical rules and constructions, reveals the calculative essence of the rational logos formalized by Aristotle. The mathematization of empirical cognition of nature in modern times reveals the a priori casting of observable nature, even as an electromagnetic field, to be essentially continuous with the nature represented as a multiplicity of substances whose characteristics inhere in them as properties, exchangeable in their places, by Aristotle.

In what kind of apprehension of the phenomenal field are we subject to an imperative that first casts the phenomenal field in the substantive form? In what sphere can we locate the imperative which both requires us to apprehend the patterns of the phenomenal field as substances with properties and requires our concepts and judgments about them to have the predicative form?

This sphere Heidegger identifies as the practical sphere. We apprehend phenomenal entities as substances with properties when we envision them not as natures emerging in nature nor as artworks but as implements. And when we envision implements, that is, useful structures, not from the point of view of the user but from the point of view of the maker. Substances with properties are formulated in predicative judgments. Aristotle's physics and logic was a physics and logic of the practicable world.

Contemporary physics finds the natural universe cast a priori before the observer not as a multiplicity of substantive objects (*Gegenstanden*), nor as an amorphous medley of sense, but as a self-contained system of energy units which maintains itself by unending transformations in everywhere determined ways, a fund or stock (*Gestand*) of energy-units whose determined patterns of transformation deliver them over to calculation. The transformations are not represented as transformations that realize a *telos*, not represented as transformations into the better or the worse, the more valuable or the less valuable; they are represented as transformations into the equivalent. The

228

mathematized observations of these transformations calculate them in equations. The thought that formulates this universe takes all entities to consist not in subsistent substances but in quanta of force in actual or potential, but always determinate, transformation. The relational logic which regulates it calculates these transformations. For Heidegger, the shift from a substantive perception of nature to an objective observation of a world cast as a world of facts and to the projection of the universe as an electromagnetic field are avatars of an aboriginal casting of nature into a calculative form; the shift from a predicative logic to a relational logic is an avatar of the aboriginal casting of thought into a rational, calculative form.

Technology is the effective transformation of the resources of nature into products. Technological transformation is not regulated by a *telos* of the better, the more beautiful, the more sublime– the incalculable or by the uses and needs of men, for technology extends into human and genetic engineering, dealing with human nature as a fund of psychic and biological resources subject to transformation in turn. Technology which seems to be available to any purpose assigned, transforms purposes and those who assign purposes. Technology destines all its products and its purposes in turn to be transformed into the equivalent; it is a process of transformation that has no finality and no end.

In our technological age man is *Homo faber*. More and more the environment that surrounds us is manmade. Our research laboratories do not study natural entities, but pure water, pure sulphur, pure uranium which are found nowhere in nature, which are produced in the laboratory. The table of elements itself is no longer an inventory of irreducible physical nature; atomic fission and fusion makes them all subject to transformation. It is not its own nature, its properties linking it with its natural setting, that makes a thing useful to us, but the properties it reveals when inserted into the instrumental system we have laid out. Timber is first cut into rectangular boards before it can be useful; the trees themselves are first hybridized, thinned out and pruned, before they can become useful as timber. It is not willow bark in its nature as willow bark that we find useful for our headaches, but the extracted and purified essence synthesized into aspirin tablets. There are whole plantations now where biologically engineered species of plants grow not on the earth but in water, anchored on floats of plastic foam fed by chemical blends. There are reserves now where genetic engineering is producing new species of patented plants and animals. As a biological species we are ourselves a feedback product; our specific biological traits, our enormously enlarged neocortex, the complexity of our bodies' neural organization, the expanded representation of the thumb on our cortex, our upright posture, our hairlessness did not evolve to differentiate us from the other primates naturally, but evolved as a result of our invention of symbolic systems, evolved from feedback from the culture – the perfecting of tools, the organization of hunting and gathering, of the family, the control of fire, and especially the reliance upon systems of significant symbols – language, ritual and art – for orientation, communication and self-control. As a bionic species we find these systems of communication subject to the imperative form of calculative reason.

Across the planet all nontechnological cultures collapse before the power and the dominion of Western or technological culture; in the West festivity, art, and religion are transformed into technological dreams and dances. The historical imperative of the technological epoch makes itself universal and necessary. The Rhine of Hölderlin's *Homecoming* has become a hydroelectric power plant; the festive chalice that poured forth wine and blessings to many has become a collector's investment; the heavens have been chartered with Riemannian geometry and their resources calculated; the music of the spheres is synthesized; art has ceased to be a locus where the truth is revealed and has become a stimulus for the technological will to power; the gods have fled before unendingly metamorphosing holograms projected by laser beams; birth is genetically engineered and perception conducted through electronic microscopes and telescopes and feelings biochemically adjusted and the dying of individuals in intensive care units as of populations technologically administered. Heidegger's insight into the technological imperative which weighs on our history becomes an apocalyptic vision of the end of history in an endless recurrence of the same. Heidegger, more yet than Kant, sees in nature, in the practical field, and in the social field one and the same form of order; sees in the ever more multiple partitioning of scientific disciplines, as in the ever more narrowly focused directions of technological invention and in the ever more rapid evolutions and revolutions of social organization the inescapable dominion of one form of order. Which imperiously commands that man make himself a pure fund of physical and psychic energies subject to transformation without ends and without end.

The mathematization of nature, the casting of nature into the form of a fund of material subject to indefinite technological transformation into the equivalent, is not simply a willful project of a natural species conceived for its own uses and needs. The casting of thought in the form of reason is not a program willfully fixed by ourselves. The casting is a destination, an imperative laid on our history. Heidegger's inquiries into other world-epochs, in which other determinations of the practicable field and of the perceived world and of the human community and other forms of thought prevailed, do not imply that their laws could be also our law, or that we could freely select which form of law to follow. Their real purpose is to bring out more forcefully into relief the exact nature of the logico-mathematical, rational, finally and originally technological, imperative to which we are subject. In order that thought of a logico-mathematical form, thought in the form of reason, arise it was necessary that the content lend itself to being apprehended to this form of thought. The effective transformation – which finds no limit, which in an atomic age is now driven with an energy-source in principle without limit – of nature into a pure fund of products open to indefinite transformation into equivalent products, is the evidence that the world lent itself without reserve to being apprehended by this kind of thought. It was then being itself that presented itself in such a way as to impose and require this kind of thought; it was being itself that exhibited the sensory patterns in such a form as to lend themselves to be apprehended by calculative reason.

The segmentation of the theoretical field

And yet, when we examine the history of our Western scientific representation of nature since Newton, our applied or technological science, and our representation of society and culture, do we not find that our science undergoes periodic fragmentation of its domain through the introduction of new technologies for observation and verification in particular empirical fields, but also though the introduction of conceptual systems and of models and types of laws specific to particular fields of data? And perhaps new and essentially incomparable forms of technological transformation and nontechnological transformation and nontransformation.

The representations of the instrumental organization of material reality or the structures of society our sciences furnish us with are not constructed around the same kind of formal laws that regulate our scientific representation of nature.

Technology is not simply applied physics, nor reified mathematics. The Heideggerian argument that the substances physical science observes are technologically produced does not demonstrate the isomorphism of logico-mathematical theory and rules of practice. The practical properties of a substance are not simply revealed by its physico-chemical analysis; they are revealed only when related to the practical end, which in turn is not determined by its physical properties and physical dynamics but by its position within a technically arrayed order. It is true that technology draws on physics and returns back into it, in the sense that the substances that serve as the data for empirical observation are not natural but technologically produced substances. But the rules for the technological production of those substances are not of the same form as the physical laws of the substances out of which the technological equipment is produced.

Kant had taken the representation elaborated in jurisprudence to be the representation of the specific kind of order, in relations of command and obedience, which constitutes the multiplicity of individuals in interaction into a society. He had assimilated the universality and necessity of the laws with which the rational jurisprudence of the Enlightenment represented the order that constitutes civil society with the formal properties of the empirical laws that constitute the representation of empirical observations into a representation of nature. But in reality the jurisprudence that formulates juridical practice employs a distinctive form of reason, which does not elaborate the universal and necessary rules of social intercourse and deduce juridical judgments from them, but argues from cases and precedents, with a normative reason which is not simply that of empirical generalization. The jurisprudence of enlightened despotism set out to recast the jurisprudence based on monarchic decrees into the forms of natural science. Michel Foucault[2] has pointed out that this monarchic-discursive representation of power was elaborated by theorists of enlightened despotism at the very time when the monarchic-discursive organization of power was giving place, in Europe, to power-relations seated in the archipelago of new disciplinary institutions – barracks, factories, schools, hospitals, asylums. The bourgeois revolutions dismantled the monarchic structures of power, and its political theorists enlisted mass support for this enterprise by proclaiming as their own the program of the jurisprudence of the Enlightenment, a monarchic-discursive reorganization of power as the subjection of every individual citizen to the universal rule of rational law. But in reality the effective regulation of each citizen in modern mass-democratic societies is realized not by legislative decree but by the disciplinary technologies elaborated in schools, barracks, factories, hospitals and asylums. The monarchic-discursive representation of the power structure in society then did not represent theoretically the functioning of power in European society; it functioned programmatically to dissimulate the new forms of a social order, which, like every form of power, is effective in the measure that it masks a good deal of its functioning. Contemporary jurisprudence recognizes that its forms of reason are not those of natural science. It advances not by integrating the juridical decisions made by arguing from case and precedent under universal and necessary rational laws, but by integrating into the body of legislative decrees the rules of the disciplinary archipelago. And it too functions to mask a good deal of the effective relations of power with which modern society is constituted.

For contemporary social science jurisprudence ceases to be a representation organized internally by laws of the same kind as those in effect in the natural sciences, and which represents the laws that constitute civil society; jurisprudence itself fragments. And jurisprudence which Kant took to represent the intrinsic order of civil society does not reveal the social domain, which classical science since Aristotle had been taken to comprise the family organization and an organized economy as well as a political organization as

232

civil society.

Contemporary social science has constituted as separate regional domains multiplicities of cultural observations that had not even been included in the empirical domain of classical social science. New segments of the social and economic sciences have organized the specific regularities that organize into intelligible domains the monetary order, human genetics, the kinship relations of gender-based societies, the management of human resources, the structure of games, linguistic pragmatics and kinesics, the long-range field of the operation and structural transformation of myths and the short-range field of media-generated representations. The generalized uneasiness about the scientific status of these cognitive specializations is less an uneasiness about the power of their laws and theories to organize the domain of the data than an epistemological uneasiness about the internal fragmentation of the domain of theory. Psychoanalysis has made its real advances when it abandoned the neurological and behaviorist concepts that would have integrated the data of the psychic sphere into the domain of general psychophysiology. Cognitive science has made its advances by abandoning interpretative concepts and schemata from general psychology and contriving regional concepts and hypotheses. Linguistics was the first of the cultural sciences to have achieved a scientific status, and it did so because the concepts of phonetics, syntactics, semantics, and pragmatics are specifically regional, and the paradigmatic rules it has been able to formulate for the structures and evolution of these strata of a language do not have the form of formulas with variables and rules of combination and transposition.

Certainly the multiplication of disciplines has engendered theories that have tried to make connections to integrate the disciplines into more general theories. But the very success of these general theories engenders new disciplines outside of them. Structural linguistics was born of the introduction of the economic concept of value alongside of the semantic concept of a term. The functioning of a linguistic term is envisioned within the structure of an exchange-system among terms and not only within the structure of an exchange of representational signs for referents. The values of the significant units of the structured strata of language, the syntactical and grammatical, logical and semantic rules for their distribution and circulation, are made intelligible within the specific limits of the specific kinds of structures within which they function and within which they evolve in systematic ways. The strata are not isomorphic; the phonetic system of a language evolves systematically but does not evolve as its syntax evolves. The diachronic analysis of language will, however, also be structural; there is a system to the relationship between the phonetics and the grammar of a language and a system to the transformations the evolution of the one induces in the other; a general linguistics will be attempted. Social facts in general are significant

233

facts, facts functioning in virtue of their meaning, and functioning within structures of exchange which maintain themselves and transform themselves systematically. From the start Fernand de Saussure conceived of and Claude Lévi-Strauss set out to realize a generalized semiology. The Aristotelian analysis of the social order is now envisioned by Lévi-Strauss as non-isomorphic systems of circulation of goods, women (the Aristotelian fundamental unit of society, the kinship relations that constitute the family, defined by Lévi-Strauss as a system of rules established by men for the distribution of women), and messages. Indeed the economy of a single region, such as the Mediterranean basin under Charles V, fragments into distinct evolutions of diverse productive and mercantilist economies; the technological system organizes and evolves differently from the distribution system; the English language fragments into an open-ended multiplicity of class, professional, regional, and subculture languages. Lévi-Strauss elaborates structural analyses of the taxonomies and syntax of different cuisines, of the different fashions that form and transform in contemporary societies, of the garb and traditions of the Académie Française which is charged with maintaining the integrity of the French national language, and his analyses are not simply derivations from the universal logic of thought or from the finite system from which an indefinite array of information systems can be generated. But the multiple, overlapping, intersecting, and divergent systems transforming each according to its own tempo transform and disintegrate in systematic ways, and interact on one another in systematic ways. Structuralism arises as the general theory of the synchronic and diachronic structural relations between the multiple structured strata of society. The very success of generalized structuralism gave rise to new general theories extending information-theory, linguistic pragmatics, generative semantics, game theory, cybernetics, aimed at integrating under one form of intelligibility the multiple systems of forms of laws specific to regions of culture, and which extend, distend, and fragment ever further the domain of the forms of scientific thought.

In natural science itself, we see not only a prodigious augmentation in the accumulation of observations and a multiplication of theories but a multiplication of the forms of theories and a fragmentation of the theoretical order. Not only do empirical laws have no true necessity, but statistical laws are instruments for the deduction not of determinate eventualities but of probabilities which may even in principle escape precise where-when determination. In quantum mechanics one cannot specify simultaneously the position and the velocity of any elementary ultimate particular; any conceivable technology to establish the observation of one coordinate renders the determination of the other indeterminable. It is true that quantum field theory is based on the theoretical project of unification of relativity, quantum mechanics, and cosmology. But the concept of universality as the ruling value for a theory has become problematical. It is true that any new conceptual scheme, model, or method for elaborating theories introduced is transferred to heterogenous domains and produces theoretical renewal and advance. But in these transfers, concepts introduced into other systems become metaphorical, laws acquire different powers of extension and different meanings, paradigms reorganize but also restrict the phenomenal field accessible to investigation. Solid state and particle physics do not constitute themselves simply as separate regional disciplines in Husserl's sense, differentiated by their fundamental concepts that delimit a specific region of data; they differ too by the forms of their laws. Indeed this was already the case in classical physics. It is true that the Standard Model being elaborated seeks to integrate quantum physics, relativity theory, and cosmology. Yet Kuhn, Lakatos, Feyerabend, Fleck, Laudan, Hacking, van Fraassen, and others have argued that the most productive scientific advances made in recent times have worked with hypotheses and theories that are essentially regional. It was by fragmenting the unity of Newtonian physics with the macrocosm-specific theory of relativity and the microcosm-specific theory of quantum mechanics that twentieth-century physics realized its greatest intellectual achievements.

We do not have the theories of translatability that would integrate microbiology into genetics, neurology into biology, biology into physics, quantum physics into relativity theory. The theories that make the connections between the laws of one discipline and those of another themselves fragment into new disciplines without univocal translatability. These theories do not even have the same mathematical form; mathematics itself has lost its unity and fragments into region-specific mathematical disciplines. We do no have one many-branched empirical science dealing with the natural universe; we no longer have an effective concept of the universe, even as a regulative ideal. The universe is the specific topic of a cosmology that does not have the totality of the empirical data of all the other scientific disciplines as the basis of its laws and theories, that indeed elaborates its laws and theories with the least data available or possible. We have many non-reciprocally translatable scientific disciplines with neither unity of conceptual system, logic, mathematics, or method. We have an array of actual and probabilistic empirical regions without the universal and necessary laws to integrate them into that one representation that would be nature.

It is argued that the fragmentation of the domain of natural science in the measure that scientific theory advances is not provisional; it has causes that are intrinsic.

The empirical laws of natural science are not universal and necessary principles from which the observation-reports can be shown to be deducible. Empirical laws are not invalidated simply by counterinstances. There is, Paul Feyerabend claims, today in physics no law without exceptions, not a single theory that covers all the observed data of the relevant sector it formulates.

Observation-reports do not give facts that verify or invalidate hypotheses. For science knows no "bare facts" at all, but the "facts" that enter our knowledge are viewed with certain assumptions about the physiology of the perception taken as normal, identified by certain theories about what factors can be disregarded as insignificant when one isolates one item from the rest of the electromagnetic fields of the universe, reported with certain theories about the definitions of the concepts with which they are formulated and which presuppose a certain logic, recorded in measurements that presuppose the theories of a certain mathematics.[3] The practice of telescopic observation and acquaintance with the new telescopic reports in Galileo's day changed not only what was seen through the telescope, Feyerabend notes, but also what was seen with the naked eye.[4]

Right method in science cannot consist in choosing between theories which have been tested and those which are falsified because there is no such thing as a crucial experiment in empirical science. Since one never isolates any reaction from the rest of the universe and never isolates any theory from the body of theories connected with it, the experiment that does not give the result the hypothesis had predicted does not determine whether the fault lies in interferences from the surrounding universe, in which of the complex of interinvolved theories being tested, in the mathematical system involved in the measurement, or in the logical system involved in formulating the hypothesis and in reporting on the experiment.

Consistency of a law with the other (always relatively) confirmed laws of the (already theoretically shaped) facts of the scientific discipline cannot be a criterion for the validity of a empirical theory. No advancing science shows this kind of consistency. Newton's theory is inconsistent with Galileo's law of free fall and with Kepler's laws; statistical thermodynamics is inconsistent with the second law of the phenomenological theory, wave optics is inconsistent with geometrical optics, etc.

If the ad hoc character of many of the most important theories in natural science does not invalidate them, this shows that theory-building in empirical science does not obey an aesthetic imperative of unity and simplicity. Nor is theory-building pursued, as Kant and Husserl reasoned, as the means by which the theoretical subjectivity comprehensively and integrally maintains all its own past and future phases and states, in obedience to an imperative that commands it to maintain itself as an abiding and selfsame, ideal, presence. If the counterinstances found to all empirical laws, and the unlocatability, by crucial experiment, of the fault in a complex of theories or in the logic of the forms with which that theory-complex was constructed and which enables us to make experiments do not invalidate laws and theories, this is because the most useful function of a new hypothesis, a new theory, is not to account for hitherto

236

unexplained and unpredictable facts. The geometrical model of theoretical science as a method to engender all possible laws and all possible empirical observations deductively is not the ideal of but a misconstruction of empirical science. The function of empirical theories is empirical; useful empirical theories generate new facts. Galileo's theory prevailed not because it simplified the organization of existing astronomical observations and rendered the calculations for new observations possible, but more importantly because, through its inconsistency with optics and the physiology of perception, it generated the production of new theories in these areas, which generated new kinds of observations. Because the value of laws and theories does not lie in their formal properties, it follows, Feyerabend concludes, that there is no way to methodically distinguish nonscientific from scientific concepts and theories; ancient acupuncture diagrams and treatises of Ayurvedic medicine are no longer a priori disqualified by competent scientific researchers; theories and paradigms and conceptual schemata from narratives, myths, theologies, literature, dramatic and lyrical arts can well prove effective in relating known observations and turning up new observations.[5] The empirical domain for contemporary science is not that cast in advance in the form of units of energy in transformation without ends and without end.

Phenomenology of the ordinance

Scientific as well as nonscientific, mythical and theological theories, tradition and literature are second-order representations; theories consolidate with an encompassing order the perception that apprehends sensory patterns as things and landscapes, the practical view of our sensory-motor organisms that envisions the force or obstacle action encounters in a broader layout of means and ends, and the pledges, pacts, alliances, compromises, and armistices with which passionate men envision the social field. The theoretical representation of nature does not disqualify and replace the perceptual discovery of the world; geology derives from and presupposes not only scientifically conducted explorations but the inhabitation of the planet by humans in the multiple concerns of their lives; biology and botany and pharmacology and astronomy working with scientifically conducted observation presuppose the natural perception that locates animals, vegetables, minerals and celestial bodies and locates for the scientific researcher his laboratory and his implements and his books. Technological operations still and always require the natural sensory motility of the skillful body of the one that leaves his bed and his home for the laboratory and the workplace. The mathematical calculations with which Lévi-Strauss represents the elementary and composite relations of kinship of gender-based societies presuppose alliances and recognitions of descendence with which those societies form, with which individuals associate without having calculated their forms of association mathematically, and associate with the anthropologist who comes to them.

Theory does not order an amorphous content diversely gathered by perception, by manipulation, and by social encounter; there is already an ordinance in the field of perception, in the range of action, and in the arena of social interaction. The observations of the natural scientist, equipped with precision instruments and focused by a research program, are oriented within the layout of the world given in natural perception. Technological engineering theory has to start from and return to the carpentry of practicable reality open to human organisms and their skills. Social theorists seek with their conceptual tools to formulate an ordinance that takes form when men associate without those conceptual tools.

What are taken to be purely rational ideals and imperatives – the universality and necessity of laws that would compose the patterns of sensation into one nature, unity and simplicity in the forms of our representations, logico-mathematical calculability – are themselves dictated by the ordinance of the perceived world, practicable reality, the social field. Lévi-Strauss founds the deep structure of scientific as well as savage reason on the microcosmic structure of living things and perhaps of matter, Whitehead founds the intellectual ideal for unity and simplicity on the cosmological imperative for order, but the scientific representation of microcosmic or macrocosmic reality is itself commanded by the world unfolding in perception. Kantism founds the intellectual imperative of universality and necessity on an ethical imperative to act as a responsible citizen of the universe; Heidegger founds the rational imperative for calculability on a technological imperative for transformation of practicable reality. The representation of the universe which can only be a collective enterprise must find its imperative in the ordinance that commands the formation of the social field. Ad hoc theories in natural science, in technology, and in social science derive their authority to displace the order of the existing theory, to rend the consistency and coherence of a body of theories, from the ordinance of the perceived world, of the practicable field, and of the field of social interaction.

The layout of the field of perception functions as an ordinance imperative for perceptual exploration but also for theoretical organization of observations; the paths between manipulable things function as an ordinance imperative for action and also for technological theory; the lines of action-at-a-distance between oneself and the appeals and demands of which others are the loci functions as an ordinance imperative for our forms of association and also for social science. To understand not only the form scientific theory gives to nature, to the practicable

238

reality, and to civil society, but also the imperative character of that form, we have to bring forth the ordinance in these first-order presentations.

While the world of perception is the source of all scientific data and all its verifications, the scientific representation of nature also alters the world of perception. It not only provides us with new prosthetic organs for perception; the ways it shapes the social and practical ordinances shift the levels of the perceptual world. The world we perceive is not a nonhistorical transcultural constant. Technological science does not only introduce new instruments and new obstacles into the practical field. While technological science derives its imperative from the layout of practicable reality, the perceptual ordinance that regulates its observations and the social ordinance that commands its organization enter into the structure of practical reality unfolding before our skills and our manipulations. And while the ordinance that commands human association in the passion for wealth, for power, for prestige, also commands the representations of social science, in turn economics, political science, and social psychology commanded by perceptual and practical imperatives inevitably reorient and do not only mirror the ordinance of the societies they depict. The presence there of the anthropologist (obeying the ordinance of his own society, committed to observation and to a perceptual imperative) distends the economics, the political forces, and the kinship relations of the hitherto isolated societies he observes.

Because the perceptual world is not an amorphous sensuous flux, it is open to a phenomenology of perception that would describe the array of things as apprehended by natural perception and not only the organization of objects as observed with the precision instruments and research programs of natural science. Practicable reality is open to a phenomenology that would describe the field of action as discovered by our mobile and manipulative organisms. The social field is open to a phenomenology that would describe the encounter with others as loci of appeals and demands and the associations formed with them. If it is in logic and the history of scientific systems that we seek the account of the order that regulates our theoretical representations, it is in descriptive phenomenology that we seek to delineate the ordinance that reigns in the layout of the world of perception, in the field of action, and in the field of social interaction.

It is only this kind of descriptive phenomenology that can lead us to understand the imperative ordinance that commands the forms of our theoretical representations of nature, of practicable reality, and of the social field, but also alone makes it possible to identify those forms. And to identify their entry into the layout of the perceptual world, the range of practicable reality, and the social field as historical imperatives.

239

The order of products; the ordinance of implements

Heidegger, we have seen, proceeds by disengaging, from formalized mathematical logic, the calculative essence of rationality. Rational thought is thought a priori cast in the form of propositional calculus. Rational thought, which gives itself out as speculative reason, thought that represents the given, is possible only on the basis of a prior casting of being in a calculable format. The original locus of this casting of thought into the predicative form and of the given in a substantive form Heidegger locates in praxis. He identifies it as thought that formulates the experience of useful things, envisioned from the point of view of the maker. Heidegger thus proceeds by putting into relation the form of representation with a prior presentation. The order of things represented, in predicative discourse, as substances given over to calculative thought, derives from the ordinance of things cast as products in the practical experience of man cast as *Homo faber*. Predicative thought becomes scientific in being recast into its integrally rational, logico-mathematical form; the world of substances is recast into a fund of energy-units whose transformations lend themselves to calculation. The scientifico-technological representation of the universe continually alters the field of production, not only providing it with new technological means but also with new materials, new forms, and continually transforming the finalities first produced. This casting is a historical event, not simply an event locatable in time, but an imperative that engenders a history, a technological history of the unending transformation into the equivalent.

But the prior representation of a field of implements to the maker is itself a historical event. The order of things cast as products derives in turn its ordinance from the things cast as implements for man the user, *Dasein* as project. This is not to say that there is an aboriginal, ahistorical and transcultural, natural world given as implements to a human nature which exists as prehension of means in view of ends. While the field of products derives from its ordinance, Heidegger will argue: from the world of usage, the transformation of the world of production under the effect of the technological representation of the universe solicits and occults, and continually transforms the world of usage and the user.

The point of view of the producer, the creator or maker, is, Heidegger claims, in fact derivative.[6] it is the user, the usage, that apprehends the essence of things that are because they are useful. If you want to understand shoes, ask a dancer, ask a runner, ask a mountaineer. The shoemaker himself must ask them. If you want to understand a motorcycle, ask a biker.

For the user to deal with an implement is to envision a certain practicable field, in which it maintains and orients his position, sustains and facilitates his advance. When the hiker reaches for her boots, her view does not stop on the shape of a mass of leather; she turns to the mountains, the crisp air and damp and pellucid light, the rocks that rise over her as challenges and as sovereign panoramas over those that circulate in the plains. Her boots are not for her raw material but a supporting form sustaining the forms of her own advance, the power to forget the mud and the thorns. In the materiality of the boots, she sees the consistency, the reliability that sustains the rigor, the harshness of mountain paths. The shape of the boots is not for her that geometrical diagram which guided the shears of the shoemaker but a form that corresponds to the contours of the rocks over which mountain goats leap and eagles soar and of the hollows of the brush-covered resting places of deer. In the form of the boots she sees the orientation of her forces towards those paths, in their materiality she sees those paths becoming practicable. The boots belong to the mountains, and make her belong in mountains. It is for his neighbor, the spectator, that his motorcycle is so much metal of a certain shape cluttering the view into the back yard or obstructing the driveway; for the biker, his bike is power, speed, roar, phallic thrust that he feels in the confidence and assurance of his body weaving through traffic, plunging by the semis on interstate highways, in a world become a maelstrom of burning rays of sun and roaring winds and nights in which all small lights of rooms about which the sedentary murmur before television screens hurl by like meteors.

The implement, then, for the user, is not simple means, intermediary between the image of his needs he produces in his mind and their satisfaction, link in a circuit that goes from himself to himself. It is her anchorage where the substantiality of the terrestrial which supports all things in their places supports her movements toward them, and the focal points of axes along which a practicable field – the dance floor, the mountains, the open roads – extends before her. She does not view the materiality of the implement as a mass of resistance closed in itself; her movement measures it as a contour in the supportive substantiality of the terrestrial. She does not view the form of the implement as a closure which functions to maintain the thing in its own place and time and exclude the rest from itself; the form of the hiking boots exist on the mountain path; their consistency, their durability, their reliability, their flexibility, the affinity of their leather for the warmth and the damp of the earth is felt in the hike; their firm shape, their lines of tension which are lines of support, exist and are known in the rocks and the cliffs. The solidity of the motorcycle is revealed on the dirt trails; its streamlined form revealed in the gale winds bringing the Pacific Coast down interstate highways.

For the user, things are not substance formed by properties which make those things properties of the producers; the world of the user is not formulated in the predicative logic which operates with particular subject-terms which are not predicated of other terms and to which universal predicates are ascribed. The relational logic which Whitehead saw as regulating the more advanced scientific representation of objectivity, depicting the universe dynamically as an electromagnetic field for an operative and no longer contemplative conception of theoretical praxis, Heidegger finds to be required by the original format of the world, practicable field for the user, who is himself constituted as a user by being a concrete nexus of prehensions, concrete enactments of relatedness.[7]

Yet is it not true that an implement exists in view of an end, that it is the end that makes the means? And is it not true that the end, the final cause, is an idea, a mental diagram first produced in the mind of the maker, with which he initiates his production of useful products? Does not the fact that the maker first makes the finality make his point of view on the implement privileged?

The primary discovery of reality, practicable reality, in hiking, dancing, biking, feasting, this primary relational format of finalities, of implements, and of the forms of openness materialized in them gets converted into that other perspective from the point of view of the maker, for whom finalities are mental images he first makes, and for whom implements are substances he makes by imposing forms on chunks of raw material. The privilege of this point of view, the primacy of *Homo faber*, is represented in Western culture as something destined; it is man seen in the image and likeness of his creator, that is, of God destined to be visualized in the image and likeness of a maker. The privilege of this point of view is not simply elaborated in theological representations; it is realized in our technological world-historical epoch which finds all things subject to measurement, whose thought, essentially rational, calculative, conceives of all things known and knowable as a pure fund of energy-units subject to transformations in determinate ways, whose empirical research observes all things in and through transformations. And which advances across the planet and into outer space technologically transforming nature into a fund of resources without finality subject to transformations into the equivalent without end.

The ascendancy of the point of view of the maker over that of the user is realized in the replacement of implements with consumer goods. The fabricating of an array of implements whose consistency reliably supports the hiker's, the dancer's, the celebrant's place and movements in the fields and finalities that summon him to the exterior is replaced by the production of consumer goods, products made for consumption. What determines the consumer products to be produced is the calculation of their salability. Utility is replaced by profitability, reckoned mathematically. The sales price to the consumer includes in the cost of the technological transformation of the material into the product the cost of the technological transformation of images of finalities in the mind of the consumer.

This does not mean that the modern industrial world is engendered by a Copernican revolution which locates the law of production no longer in the external finalities for which implements are useful, but not in the will of the producer himself. The designers, the decision-makers, the manufacturers do not select the products to be produced by consulting the will of the users or their own taste and values; the decisions are made by marketing research calculations of what cost-effective transformations can be made in the current directions of consumption. Producers survive only if they can continually transform their line of products with the season and the market, and continually transform their own selecting, decision-making, organizing function. Both consumers and producers obey an imperative to transform all things into products and all products into the equivalent.

In assuming that the technological imperative has come to prevail in the world of perception, which Heidegger takes to be a derivative region of the primary world of praxis, and in the social field, understood now by rational thought and subject to transformation without end and without ends, Heidegger disengages an imperative unification. For if indeed rational thought and technological praxis enter into the social field, reorienting the ordinance with which passionate men associate, reorienting and indefinitely extending the arena of the wealth, power, and prestige, it is also true that rational thought itself and the technological production it informs can only be a collective formation. The scientific representation of the universe can only subsist in the social institutions. It responds not only to an ordinance in the practible field of producers and in turn that of users, but to the ordinance which commands the association of men. The multitude of men do not only figure as psychophysical energy-units delivered over to technological transformation, but as a multitude of loci of appeal and demand which are associated in a social imperative.

Likewise, if indeed rational thought and technological praxis enter into the world of perception, recasting natural perception in the forms of scientific observation, reorienting the levels of the perceived world into dimensions of a fund of resources to be transformed, it is also true that technological products, as implements before them, take their place in the landscapes of perception. It is still the landscapes and levels of perceptible reality that give their index of reality to the technological products and their rational forms. Radio telescopes, scanning electron microscopes, stethoscopes, ultrasounds, CAT-scans, radioisotope tracers, DNA sequencers, and other prostheses have populated the real world of perception with myriads of new entities. But it is middle-sized entities, susceptible to direct or indirect perception, that have succeeded in attaining the status of reality; those too great or too small retain the status of conceptual entities or instruments of calculation. It is the ordinance of the perceivable world that shifts conceptual entities into the density of reality.

Notes

1. [*The author originally submitted a slightly longer version of this essay as one of two for inclusion in a separate collection edited by the first editor: Babich, B.E. (1995), <u>From Phenomenology to Thought, Errancy, and Desire: Essays in Honor of William J. Richardson, S.J.</u>, Kluwer, Dordrecht. One essay was published in the aforenamed Festschrift and because of this essay's topical relevance the editor requested, and the author most graciously agreed to permit, publication in the present volume. – Ed.*]

2. Foucault, M. (1980), *The History of Sexuality, Vol. I: An Introduction*, Hurley, R. (trans.), Vintage, New York, pp. 81-91; and *Discipline and Punish*, Sheridan, A. (trans.), Vintage, New York, Part III.

3. See Putnam, H. (1990), *Realism with a Human Face*, Conant, J. (ed.), Harvard University Press, Cambridge, pp. 115-117.

4. Feyerabend, P. (1978), *Against Method*, Verso, London, p. 132.

5. We should take into account how theological treatises are used as instruments for the discovery of logical problems and procedures, how literary genres are used as organizing schemata for historical narratives, how rhetorical tropes are used as organizing models for political economics (White, H. [1973], *Metahistory*, Johns Hopkins University Press, Baltimore), how psychoanalysis uses myths as diagrams for complexes, how musical models are used as diagrams for understanding myths (Lévi-Strauss, C. [1981], *The Naked Man*, Weightman, J. & D. (trans.), Harper & Row, New York, pp. 646-67.)

6. God conceived in the image and likeness of a maker, God the Creator, is derived from the gods invoked as responsive to earth and skies and man's mortality in a relational field of which the usage-object, the jug, is the concretion. Heidegger, M. (1971), *Poetry, Language, Thought*, Hofstadter, A. (trans.), Harper & Row, New York, pp. 166-74.

7. Heidegger will thus describe the implement as used not as a substrate with its own properties but as a nexus of "concrete modes of appropriateness"; he will describe the user not as a subject with attributes issuing movements but as a projective, ec-static power, comprehending and affectively disposed. The user does not have the factual being of an entity that is where and when it is but ex-sists toward

244

the beyond and the future. But Heidegger does not examine the relationship between the relational discourse Whitehead has identified as required by contemporary physics.

14 Eros, Thanatos: The Emerging Body in a Postmodern Psychology of Science

Brian Pronger

I would like to embark on a reflection, a reflection on the eventful experience of the body and the status of that experience in the psyche. Within that context I will consider the place of science. I will then suggest how these reflections constitute a post-modern psychology of science.

It is essential that this thinking about the body and the psyche be grounded in the bodily experience of being-there oneself. I invite you, therefore, to read these words not in abstraction but in the context of your own eventful experience. What truth I may allude to here will be such only in so far as it frees one for that experience. And so I ask you to reflect on a time when you have had a heightened awareness of the event of your own bodily experience, the event of your own being-there. You might reflect on a period from a run you had last week, a period in which you were aware of the ecstasy or the agony of the moving body. You might remember the occasion when you and your friend were pitching and hitting a baseball, and there was a perfect moment as the bat and ball met with a thwack and you were one with that sweet spot in time. Or you might ponder that occasion when you were in a canoe and as you plunged your paddle in the water you knew the union of your exertion and the water's cooperative resistance. Or you may choose to contemplate a sexual experience in which the sheer presence of bodies enthralled you, or in which the exquisite pain of the orgasm remains even now the mysterious source of endless fascination. Imagine yourself being there. I suggest you take your experience as the empirical data for the following reflections.

We will take from Heidegger one thought, Being, which in his later work he named *Ereignis*. We would also be wise, right now, to heed the cautionary note from his essay "*Aletheia*": "The following remarks lead to no conclusions. They *point* toward the event [*das Ereignis*]" (Heidegger, 1943:106, emphasis mine).

These comments then, are an *invitation* to the event, which is *Ereignis*; they in no way *account* for *Ereignis*. I will interpret *Ereignis* with a neologism, "Eventfullness." Eventfullness, we will see, is the essence of being, the way of being; it is the being of being itself. Bodily experience is Eventfull.

An event is a happening, an inherently temporal spectacle. The word "event" suggests in some ways a great party, an event which no one should miss. With this word, I mean to inspire a memory of what it is like to have been at such an event. This is the feeling that you had to have been there to understand it. In fact, what the event was was being there. The suffix "-full" (with two l's) in the word "Eventfullness" is meant to reinstate the sumptuous Old English origins of it as "replete."(OED) When something is replete it needs nothing more. The suffix "-ness" in Eventfullness means the condition of an event being replete. And so "Eventfullness" means the condition of a replete happening.

What happened in the event of the body? Movement. Movement is implicit in any event. The relation of movement and event may be glimpsed if we consider the etymology of the word "event."[1] Our English word "event" comes from the Latin *eventus,* whose root is *venire,* which has its origins in the Greek root *BA* which means "go." An event goes. This is not lost to us in Modern English. For example: I might *go* for a swim. Going for a swim is not just a matter of preparing to swim, presenting myself at the pool so that the event of my swimming can take place; actually moving through the water, I am *going* swimming. The event of my swimming consists in such a going. What we derive most essentially from the Greek *BA* in our understanding of Eventfullness is the pure sense of movement. But this is not the sense of simply going away, of leaving behind. Going out of is always a coming into. If I am going into the water I am also coming into the water; these movements feel essentially the same to me. In movement the present goes from where it is present to where it was absent. When I am swimming, I go from where I am (presence) to where I previously was not (absence).

In movement, absence is penetrated by presence. Absence lies before presence, waiting. In that lying before presence, absence belongs to presence. Without absence, presence is impossible. Presence must have an absence into which it enters or it will no longer be presencing but will only have been present. Absence is present in the present as the space ahead which makes presencing possible. Presence cannot be without absence awaiting penetration.

Now the penetration of absence by presence is not an assault on absence. To truly understand penetration we must purge it of its patriarchal associations. This is *not* the penetration of an "inferior" other.[2] On the contrary, absence is equiprimordially one with presence. In fact, in penetrating absence, presence penetrates itself. Because: absence, as lying before presence, is essential to making present. That is, absence opens for presence its own being-present; as the opening for presence, absence draws presence into presence. Absence, lying before presence, equiprimordially drawing presence into presence, is inherent in presence; in this way absence is presence. Absence draws the penetration of presence into itself. In that drawing, presence penetrates itself. Drawing and penetrating, therefore, are one. Drawing-penetrating-absence-presence is the unity that is the mysterious opening of the Event that being is.

You might get a more intimate feeling for what I am saying if you reconsider the bodily experience which you have taken as your empirical reference for my words today. Think of the intense presence of that experience. Think of the opening that that experience is. Think of how you are drawn to such experiences because of what they open for you.

That opening is the freeing of presence and absence for each other; it is the freedom which absence gives presence to penetrate itself, which is the freedom presence has to repeat itself – in penetrating itself presence repeats itself. The freeing, repeating, self-opening by which presence makes itself present, is the inexhaustible mystery of being itself, the essential happening of the Event of being, which for us is the happening of the body, the time of the body. Time is the deeply mysterious, entirely gratuitous, opening of presence. For us, therefore, time is the body opening itself.

Time is fecund, because just as rich soil in spring brings forth an abundance of vegetation, so too time brings forth an abundance of presence. Why is presence abounding? Because presence, penetrating itself, is limitless. This is the limitlessness of inwardness; presence, penetrating itself, presses inwardly into itself. Such inward movement can have no limits. The body, your bodily experience, therefore, is limitless. And so the body is one with all that is. And all that is is the unity of limitless inward movement. Eros, I suggest, is the voice of that limitless unity.

Remember your own bodily experience and consider the limitless inward movement of it. Think of your erotic experience, be it running, playing ball, paddling a canoe, or having sex. Think of what speaks in that experience...

How has Eros fared in modernity? What possible post might it occupy in post-modernity? Regardless of how we delineate the pre-modern, the modern and the post-modern – and as we all know there are many debates, debates which I am going to avoid here – I think it's fair to say that deference to Eros has not characterized the predominant discourses of Western thinking, most of all of Western scientific thinking, for the last several hundred years or so. Certainly we don't hear many calling upon Eros for wisdom on how to get out of a war, how to provide social equity, or how to heal a wounded planet, or how to give relief to diseased, decaying bodies. During the march (and now, perhaps, retreat) of modernity another voice has dominated our thinking and shaped our psyche.

Eros articulates the mystery of the limitless unity of being. Whereas this other voice, which I call Thanatos because its essence is denial and negativity, this other voice addresses itself to the mastery of being. It is the mastery of being that is the pre-eminent concern of modern science and technology. The mastery of Thanatos has an important place in the human psyche, especially the modern Western psyche.[3]

Both Eros and Thanatos speak of Eventfullness, but they do so in profoundly different ways. Eros articulates Eventfullness as it shows itself, which is to say as it is *repeated* in the fecund temporality of the limitless inward movement of self-penetration. Thanatos, on the other hand, is a special translation of the Event in such a way that it is *recollected as* something else. That recollection is the product of detemporalization, or hypostatization, a powerful process of withdrawal whereby the fullness of the Event is concealed, a masterful act of will that renders the voice of Eros inarticulate.

A visual metaphor might help. Imagine a ball like a fireball. But this ball is a little different; rather than sending its flames out, it sends them in. That is, the being of this ball is its own movement into itself. That ball is Eventfullness. Eros shows the fecund temporality of the ball by its own immersion in the repetitive inward movement of the ball.

Now, Thanatos pulls out of that moving ball, as it were a thread which is then spread out across a shallow, flat plane which is linear space-time. Crucial to understanding that plane is seeing that once a time is past it is past completely and forever. True repetition, which lies at the very heart of Eventfullness and therefore being itself, is impossible in the plane of linear time. And space, being extended through this plane can never be shared (sharing would be the form of *spatial* repetition); *this* space would always be alienated from *that* space. The best that can be done in linear space-time is recollection of what is *as* here or there, has been or will be.

That linear plane is a withdrawal from the limitless inward movement, the opening of presence which is time; it is a detemporalization of the Event, a translation of the Event in such a way that it appears *as* negative absence. Now this negative absence is unlike the absence of drawing-penetrating-absence-presence, for that absence is pregnant with presence. Negative absence withdraws from the primordial unity of presence and absence. And that withdrawal is accomplished by the fragmentation or divorce of presence and absence. This fragmentation is the voice of Thanatos, the essence of human mastery.

This divorce of presence and absence creates distance and distinction. As it is articulated by Thanatos, absence is not primordially one with the occurring of presence, it is only that which presence is not. The concepts of past, present and future, here and there, are formulations of the alienation of absence and presence.

Distinctions such as this and that, fish and foul, animal, vegetable and mineral, Aryan and Arab, man and woman – the list is endless – these distinctions are drawn from the Thanatic divorce of presence and absence. That negative relation to presence, absence as *not* presence, defines the negativity of negative absence, the negativity of Thanatos.

Thanatos has been on the rise in the Western psyche for at least several thousand years. Because of its vast scope, the history of Thanatos cannot be pursued here. But I think it is possible to assert that in the modern era, Thanatos has been given more authority in the Western psyche than at any other time in human history. That authority is perhaps most evident in modern science.

Modern science is first and foremost a Thanatic discourse of mastery. Since the Enlightenment, the quest of science has been the attainment of full knowledge; it has striven to give an exhaustive account of what is, to demystify mystery, to render the mysterious as nothing more than that which is not yet fully explained. The fact that mystery constitutes the very nadir of what is, poses a deep threat to the project of modern science, indeed to the Thanatic will of the human psyche. And so modern science, the minion of Thanatos, has set out to master the mysterious.

How does such mastery come about? By the Thanatic divorce of presence and absence. That divorce is scientifically devised, for example, by the relentless imposition of distinctions and distances, and by a constant quest for determining the quantities that can arise when such impositions are made. Rendering what is by drawing distinctions, drafting distances, and calculating quantities is basic to scientific thinking. In his essay "The Age of the World Picture" Heidegger points out that science projects onto being, in advance of the experience of it, a picture of what is, thereby setting the limits to what can emerge in scientific enquiry. What I am suggesting here is that the essence of that picture is the Thanatic divorce of presence and absence, a divorce which makes possible the distinctions and spatio-temporal relations that make modern science what it is. By drawing this picture and insisting upon it, science recollects what is on its own terms. By refusing to acknowledge that which does not appear on those terms science ultimately comes to mastery. In modernity, science became the high court, the judge, jury and executioner of what is.

This human need for mastery, this power of Thanatos to divide and conquer, while affording some of us great power over ourselves, our fellow human beings, and our environment, this mastery, come to its full flower in the discourse of modern science, has finally brought us to the edge of disaster.

The negative absence of the Thanatic voice is ultimately very negative because it is the source of nihilism. The here and now, which is presence in linear space-time, are annihilated by Thanatos. They become infinitely small here and now points, points that can always be made thinner. In fact, the here

251

and now of linear space-time are impossible to experience, they are mere theoretical vantage points for the linear spatio-temporal plane. The now is so infinitely short that it is always past before it can be thought with any depth. The now, which *is* presence in the linear plane, then, is lost to experience. Moreover, the past and the future lie behind and ahead as nothing more than the negatively absent other of an inane, fleeting, untouchable, annihilated present. And spatially, the distinction between here and there involves an absurd, infinite diminution; one can always peel away more of the there in the quest for a more pure here. That infinite diminution of the here is also an infinite augmentation of the there, which is an ever expanding/imploding universe of negative absence, an assault which annihilates the fecund, pregnant promise of the unity of presence and absence.

Thanatos, particularly through its scientific discourse, offers up an impossible, indeed nihilistic here and now, the last vestiges of presence after its cruel divorce from absence. Is it any wonder that people are worried about death. With Thanatos the present cannot be experienced with any depth and the past and future are always negatively absent. In a life voiced by Thanatos we have nothing, and death seems to be the end of even that.

The ascendency of Thanatos and the authority of science in the modern era has brought us to nihilism. It is this lurking awareness of nihilism that I would suggest is the spirit of the post-modern. The modern psyche has suffered enormously from the tyranny of the almost complete domination of Thanatos, a tyranny which is evidenced by the scientization of virtually every aspect of modern life. But as Foucault points out in the Introduction to *The History of Sexuality*, where there is power there is resistance. The resistance to the overwhelming power of Thanatos, nascent though it is, the resistance to this mastery lies in the long denied voice of Eros, a voice that is now a bare whisper next to the scientific roar of Thanatos.

I believe we hear in post-modern discourse an uncanny echo of the voice of Eros. Admittedly, most post-modern thinking, with its incessant talk about simulacra, *différence,* disembodied bodies and so on, while being far from the articulations of Eros as we considered them in our own bodily experience at the beginning of this paper, there is in that post-modern thinking at least a longing for something more. I know; most post-modernists would be appalled at such an assertion; they are always eschewing as old-fashioned any hope of truth, and accept with the cynicism of the truly jaded a nomadic life of dwelling among the ruins of modernity. But even in this I think there is hope: for in much of post-modern writing there is a genuine tenor of resentment, disappointment, the feeling that we have been short-changed. That resentment I suggest echoes the voice of Eros. For what we resent is that we have lost something, been denied something. We know this loss because in the movement of our bodies we have also known something more. And so the project of post-modernism is to attend to the voice of Eros, to let the body re-emerge in the Western psyche. A post-modern philosophy of science, therefore, must appreciate the place of science in the Western psyche, a psyche which has for millennia been the struggle of Eros and Thanatos, a struggle which in modernity has favoured Thanatos, but it is a struggle which just might, now at what seems to be the eleventh hour, take a turn. Perhaps then, we will be able to speak of the post-post-modern psyche and the power of Eros.

References

Foucault, M. (1976), *The History Of Sexuality: An Introduction*, Hurley, R. (trans.), Random House, New York.
Heidegger, M. (1943), "*Aletheia*: Heraclitus Fragment B 16," in *Early Greek Thinking*, Krell, D. and Capuzzi, F. (trans.), Harper and Row, New York.
Heidegger, M. (1977), [c .1938], "The Age Of The World Picture," in *The Question Concerning Technology and Other Essays*, Lovitt, W. (trans.), Garland, New York.

Notes

1. The etymologies I suggest in this paper are not intended as scholarly historical investigations of the origins of words. These are "creative etymologies" whose purpose is rhetorical. My point is to develop for some words a background that will deepen their meaning in this paper, a meaning which is always guided, not by deference to the history of the words themselves, but by the phenomenon I am trying to describe.

2. In fact, it is a deeply patriarchal assumption that only men can penetrate. For this assumes that penetration is accomplished first and foremost by a phallus. But it is heterosexist and phallocentric to assume that the phallus is the only thing that can penetrate, or indeed, that the vagina is the only orifice for penetration.

3. I am not drawing any intentional parallel here with Freud's use of the word Thanatos regarding the so called death instinct.